Lecture Notes in Artificial Intelligence   1087

Subseries of Lecture Notes in Computer Science
Edited by J. G. Carbonell and J. Siekmann

Lecture Notes in Computer Science

Edited by G. Goos, J. Hartmanis and J. van Leeuwen

Springer
Berlin
Heidelberg
New York
Barcelona
Budapest
Hong Kong
London
Milan
Paris
Santa Clara
Singapore
Tokyo

Chengqi Zhang   Dickson Lukose (Eds.)

Distributed Artificial Intelligence

Architecture and Modelling

First Australian Workshop on DAI
Canberra, ACT, Australia, November 13, 1995
Proceedings

 Springer

Series Editors

Jaime G. Carbonell, Carnegie Mellon University, Pittsburgh, PA, USA
Jörg Siekmann, University of Saarland, Saarbrücken, Germany

Volume Editors

Chengqi Zhang
Dickson Lukose
The University of New England
Department of Mathematics, Statistics, and Computing Science
Armidale, NSW, 2351, Australia

Cataloging-in-Publication Data applied for

Die Deutsche Bibliothek - CIP-Einheitsaufnahme

Distributed artificial intelligence : architecture and modelling ;
proceedings / First Australian Workshop on DAI, Canberra,
ACT, Australia, November 13, 1995. Chengqi Zhang ; Dickson
Lukose (ed.). - Berlin ; Heidelberg ; New York ; Barcelona ;
Budapest ; Hong Kong ; London ; Milan ; Paris ; Santa Clara ;
Singapore ; Tokyo : Springer, 1996
  (Lecture notes in computer science ; Vol. 1087 : Lecture notes in
  artificial intelligence)
  ISBN 3-540-61314-5
NE: Zhang, Chengqi [Hrsg.]; Australian Workshop on DAI <1, 1995,
  Canberra>; GT

CR Subject Classification (1991): I.2, F.4.1, F.1.2, C.2.4

ISBN 3-540-61314-5 Springer-Verlag Berlin Heidelberg New York

© Springer-Verlag Berlin Heidelberg 1996
Printed in Germany

Typesetting: Camera ready by author
SPIN 10513110   06/3142 – 5 4 3 2 1 0   Printed on acid-free paper

Preface

Research on issues in the field of Distributed Artificial Intelligence (DAI) is attracting many people from around the world. Research in DAI attempts to integrate all sub-fields of Artificial Intelligence, Cognitive Science, and Mathematical Logic. These lecture notes contain the revised version of all papers which were presented in the First Australian Workshop on Distributed Artificial Intelligence. The workshop was held in Canberra, ACT, Australia on November 13, 1995, in conjunction with the Eighth Australian Joint Conference on Artificial Intelligence (AI'95). The goal of the workshop is to promote research in distributed artificial intelligence not only in Australia but also internationally. The workshop covered a wide range of issues in the field of DAI, such as formal frameworks, methodology, theory, and architecture.

We thank all the individuals and institutions who have contributed towards the success of this workshop. We thank all of the authors who submitted papers to this workshop. Our thanks are also directed to our Programme Committee members who reviewed all the papers. Special thanks are directed to one of the world's leading researchers in DAI, Professor Michael Georgeff from the Australian Artificial Intelligence Institute, who presented an exciting invited talk at the workshop. Finally, we thank the editorial staff of Springer-Verlag for publishing this contribution to the Distributed Artificial Intelligence research field.

November 1995   Chengqi Zhang and Dickson Lukose

Programme Committee

Dr Chengqi Zhang (Co-chair)
Department of Mathematics,
Statistics, and Computing Science
The University of New England
Armidale, N.S.W., 2351
AUSTRALIA
Email: chengqi@neumann.une.edu.au

Dr Dickson Lukose (Co-chair)
Department of Mathematics,
Statistics, and Computing Science
The University of New England
Armidale, N.S.W., 2351
AUSTRALIA
Email: lukose@peirce.une.edu.au

Professor Victor Lesser
Department of Computer Science
University of Massachusetts
Amherst, MA 01003
U.S.A.
Email: lesser@cs.umass.edu

Dr Anand S. Rao
Australian Artificial Intelligence Institute
171 La Trobe Street
Melbourne, Vic. 3000
AUSTRALIA
Email: anand@aaii.oz.au

Dr Toshiharu Sugawara
NTT Basic Research Labs
3-1 Wakamiya, Morinosato
Atsugi, Kanagawa 243-01
JAPAN
Email: sugawara@square.ntt.jp

Dr Nick Jennings
Department of Electronic Engineering
Queen Mary & Westfield College
University of London
Mile End Road, London E1 4NS
U.K.
Email: N.R.Jennings@qmw.ac.uk

Professor Jose Neves
Universidade do Minho
Largo do Paco
4719 Braga Codex
PORTUGAL
E-mail: jneves@di-ia.uminho.pt

Dr Keith S. Decker
The Robotics Institute
Carnegie Mellon University
5000 Forbes Ave.
Pittsburgh, PA 15213
U.S.A.
Email: decker@cs.cmu.edu

Dr Norbert Glaser
CRIN-INRIA Lorraine
615, rue jardin Botanique, BP.101
F-54602 Villers-les-Nancy Cedex
FRANCE
Email: Norbert.Glaser@loria.fr

Dr Rose Dieng
ACACIA Project, INRIA
2004 Route des Lucioles, BP 93
06902 SOPHIA-ANTIPOLIS CEDEX,
FRANCE
Email: dieng@sophia.inria.fr

Table of Contents

Definition and Application of a Comprehensive Framework for Distributed Problem Solving *

Hongxia Yang and Chengqi Zhang

Department of Mathematics, Statistics, and Computing Science
The University of New England
Armidale, NSW 2351, AUSTRALIA
Email: {hongxia | chengqi}@neumann.une.edu.au

Abstract. In the Distributed Artificial Intelligence community, the term "Distributed Problem Solving (DPS)" is widely used. However, different people refer to different types of DPS frameworks. In this paper, we define a comprehensive framework for Distributed Problem Solving. Firstly, we clarify four different types of basic DPS frameworks: DPS1 for task unique-allocation problems, DPS2 for task multi-allocation problems, DPS3 for task decomposition problems, and DPS4 for task division problems. Then, we define the comprehensive framework which can be different combinations of DPS1, DPS2, DPS3, and DPS4.

In this framework, Solution Integration (SI) is a necessary component. Three different types of Solution Integration are identified: Solution Synthesis, Solution Composition, and Solution Construction which correspond to DPS2, DPS3 and DPS4, respectively.

The definition of the four basic DPS frameworks and the establishment of the comprehensive framework for DPS will hopefully lead to a better understanding and implementation of DPS systems.

1 Introduction

In the Distributed Artificial Intelligence (DAI) community, the term "Distributed Problem Solving (DPS)" is widely used. However, different people refer to different types of DPS frameworks in their specific context, which could cause confusion in identifying different features of their DPS systems. In this paper, we will define a comprehensive framework for DPS and clarify four different types of basic frameworks. Firstly, let us review the development of DPS in recent years.

Smith and Davis first defined the framework for Distributed Problem Solving (DPS) as "the cooperative solution of problems by a decentralised and loosely coupled collection of problem solvers" and distinguished it from distributed processing in [Smith 80; Smith & Davis 81; Davis & Smith 83]. The framework for DPS is viewed as four interplaying phases: task decomposition, task allocation, sub-problem solving, and solution composition. Smith and Davis classified two

---

* This work was partially supported by the large grant from the Australian Research Council (A49530850). H. Yang is also funded by the Overseas Postgraduate Research Scholarship and The University of New England Research Scholarship, Australia.

types of cooperation in DPS: Task-sharing and Result-sharing. Negotiation is used as a metaphor of cooperation in task-sharing with Contract Net Protocol (CNP) as a negotiation approach.

While Smith and Davis used cooperative experts as a metaphor to describe the sharing of workload by breaking up the task into smaller, manageable parts and allocating them to experts using negotiation, Lesser et al. [Durfee & Lesser 87, 89] applied PGP (Partial Global Planning) as an alternative approach to result-sharing and task-sharing cooperation. PGP was developed from FA/C (Functionally Accurate/Cooperation) [Lesser & Corkill 81, 83; Lesser 91], and worked as a unified framework for different cooperation styles. FA/C and PGP approaches tried to resolve the disparity of agents' partial results, which comes from the processing of different data that agents sense at different locations, and from local, incomplete or out-dated viewpoints that agents have on the global task, by interchanging tentative, partial results [Lesser & Corkill 81,83] or partial global plans [Durfee & Lesser 87, 89] among agents.

With all of the above work concentrating on how a task is split up into smaller units and shared among a set of cooperative agents, Kornfeld & Hewitt (1981) used the metaphor of competitive as well as cooperative scientists working in the same or different areas. To explain the same phenomenon, the scientists may "hold similar opinions to one another or quite divergent opinions". Systems embedded with a scientific community metaphor are highly parallel and pluralistic. Inspired by the parallelism and pluralism features of the scientific community metaphor, two important improvements in DAI take place:

(1) It directs Distributed Problem Solving from closed DPS to open information systems [Hewitt 91] with features of non-deterministic and evolution at the system and agent level in the long run; and

(2) A hybrid and more flexible framework for DPS comes to the fore where a task can be broken up into parts (as described above), or as a whole (like several scientists working in the same area), and then allocated to one or more agents to solve.

In the later case, the systems with such a hybrid framework can achieve not only higher efficiency of system performance, but also higher fault-tolerance or reliability, and quality of solutions to the task. Such systems include cooperative VLSI design [Yang et al. 85] and engineering design [Bond 89].

In order to identify these different types of DPS techniques, we define four basic frameworks for Distributed Problem Solving followed by a definition of a comprehensive DPS framework. Solution Integration (SI) is a necessary component of DPS; however, little of the previous work has paid enough attention to the important issues of SI.

The following sections will address the framework for DPS and important issues about SI. Firstly, Section 2 defines a set of easily confused concepts, classifies four types of basic DPS frameworks and discusses the associated SI forms. Based on this section, a definition of a comprehensive framework for DPS follows in Section 3. The factors important to SI are also examined in this section. The paper concludes in Section 4 with the summary of the paper and an indication of our future research direction.

2 Basic Frameworks for Different Types of DPS

A wealth of DPS systems have been developed since the end of 1970's. How can we analyse their different features? As a first attempt, we think a detailed definition and classification of different DPS frameworks are necessary.

2.1 Definitions of Basic Concepts

We will define the following concepts precisely and use them in the definitions of different types of DPS frameworks.

Suppose for a task T, the solution to T is R. Let λ be a mapping from Δ_1 and Δ_2 to Δ_3, written as $\lambda : \Delta_1 \times \Delta_2 \mapsto \Delta_3$; i.e., objects in Δ_3 are achieved by agents in Δ_1 by applying process λ on object set Δ_2. A set of objects (agents, tasks, results, etc.) make up a Δ. We will use this formal expression to describe the concepts and DPS frameworks in this paper.

[Definition 1] **Task Allocation (TA)** is a process that allocates tasks to certain problem solving agents. The agents responsible for the task allocation make up an agent set Δ_{Ata}. Task Allocation has two styles (i.e. Task Unique-Allocation and Task Multi-Allocation) as defined below.

[Definition 2] **Task Unique-Allocation (TAu)** is a process in which a task is allocated to one and only one agent. The process can be expressed as: TAu: $\Delta_{Ata} \times \{\Delta_{Aa}, \{T\}\} \mapsto \{A_i\}$, where Δ_{Ata} is the set of task allocation agents for task T, Δ_{Aa} is the set of available agents that participate in the task unique-allocation process, and the agent who is allocated to task T after unique-allocation is a certain agent A_i, $(A_i \in \Delta_{Aa})$.

[Definition 3] **Task Multi-Allocation (TAm)** is a process in which a task is allocated by task allocation agents (Δ_{Ata}) onto more than one agent, say, n agents $\{A_1, \cdots, A_n\}$. The process can be described as: TAm: $\Delta_{Ata} \times \{\Delta_{Aa}, \{T\}\} \mapsto \{A_1, A_2, \cdots A_n\}$.

[Definition 4] **Problem Solving (PS)** is a domain-dependent computational process on a task through which agents obtain their solution for their tasks.

After the problem solving process for a task which is multi-allocated, the agents will obtain their solutions which are referred to as *feature solutions* (defined below).

[Definition 5] **A Feature-solution** is a solution to a task offered by an agent responsible for the task in Task Multi-Allocation problems (for short, f-solution). After a task is multi-allocated to a number of n agents, these responsible agents solve the same task using their own expertise and achieve a set of n feature-solutions (noted as $FS_1, FS_2, ..., FS_n$) for the task, i.e., PS: $\Delta_{Aps} \times \{T\} \mapsto \{FS_1, FS_2, ..., FS_n\}$ where Δ_{Aps} is the set of problem solving agents responsible for solving task T, or $\Delta_{Aps} = \{A_1, A_2, \cdots, A_n\}$ which is described in Definition 3.

[Definition 6] Solution Synthesis (SS) is a process that synthesises agents' feature-solutions by absorbing all the non-contradictory properties of them and resolving the contradictions in them if necessary, and achieves a more reliable, comprehensive solution to the task, i.e., SS: $\Delta_{Ass} \times \{FS_1, FS_2, \cdots, FS_n\} \mapsto \{R\}$, where Δ_{Ass} is the set of Solution Synthesis agents.

Task Multi-Allocation and Solution Synthesis are reverse processes, and they always come out in pairs.

[Definition 7] Task Decomposition (TDc) and Sub-tasks: Task Decomposition is a process in which a task is decomposed into sub-tasks where each sub-task is simpler. Suppose the task T is decomposed into k sub-tasks, $ST_1, \cdots, ST_k$, then Task Decomposition process can be described as: TDc: $\Delta_{Adc} \times \{T\} \mapsto \{ST_1, ST_2, \cdots, ST_k\}$, where Δ_{Adc} is the set of agents responsible for Task Decomposition.

[Definition 8] A Sub-solution or sub-result is a solution to a sub-task. For short, s-solution. Let the set of sub-solutions for the above sub-tasks be $\{SS_1, SS_2, \cdots, SS_k\}$. The sub-solutions are achieved by problem solving agents through their problem solving process, i.e., PS: $\Delta_{Aps} \times \{ST_1, ST_2, \cdots, ST_k\} \mapsto \{SS_1, SS_2, ..., SS_k\}$.

[Definition 9] Solution Composition (SCm) is a process that composes the sub-solutions together to form a solution to the original task, which is not the simple aggregation of them; SCm: $\Delta_{Ascm} \times \{SS_1, SS_2, \cdots, SS_k\} \mapsto \{R\}$, where Δ_{Ascm} contains the set of Solution Composition agents

Task Decomposition and Solution Composition are reverse processes, and they always come out in pairs.

[Definition 10] Task Division (TDv) and Partial-tasks: Task Division is a process that divides a task into partial-tasks with each partial-task being smaller. All of the partial-tasks form the original task. Suppose the task T is divided into m partial-tasks $\{PT_1, PT_2, \cdots, PT_m\}$, then Task Division can be noted as: TDv: $\Delta_{Adv} \times \{T\} \mapsto \{PT_1, PT_2, \cdots, PT_m\}$, where Δ_{Adv} contains the set of task division agents.

[Definition 11] A Partial-solution or partial-result is a solution to a partial-task (for short, p-solution). Let all of the partial-solutions for the above partial-tasks be $PS_1, PS_2, \cdots, PS_m$. The partial-solutions are achieved by problem solving agents through their problem solving process, i.e., PS: $\Delta_{Aps} \times \{PT_1, PT_2, \cdots, PT_m\} \mapsto \{PS_1, PS_2, \cdots, PS_m\}$, where Δ_{Aps} means the set of problem solving agents that own the partial-tasks.

[Definition 12] Solution Construction (SCn) is a process that constructs the solution to the original task by putting partial-solutions together while meeting some given constraints. It can be described as: SCn: $\Delta_{Ascn} \times \{PS_1, PS_2, \cdots, PS_m\} \mapsto \{R\}$, where Δ_{Ascn} contains the set of Solution Construction agents.

Task Division and Solution Construction are reverse processes and they always show up in pairs.

[**Definition 13**] **A unit-task** refers to any of the above task cases, i.e., a unit-task can be any task T, a sub-task ST_i, a partial-task PT_j, a sub-sub-task SST_k, or a partial-partial-sub-task $PPST_l$, and so on (sometimes, *unit-task* is shortened to *task* to refer to any of the above cases). In Definitions 2 and 3, we use the T to define *task unique-allocation* and *multi-allocation*. However, any *unit-task* could be multi-allocated or unique-allocated if applicable.

[**Definition 14**] **An Integral-task** (or local task) can be a task unit in multi-allocation problems, a sub-task in task decomposition problems, or a partial-task in task division case.

[**Definition 15**] **An Integral-solution** (or integral-result, local-solution, local-result) can be a feature-solution in multi-allocation problems, a sub-solution in task decomposition problems, or a partial-solution in task division case.

[**Definition 16**] **Solution Integration (SI)** is the process that collects and integrates a set of integral-solutions to form a comprehensive solution to the original task. It can be SS in multi-allocation case, SCm in task decomposition case, or SCn in task division case.

The explanation of all notations in this paper is listed in Table 1.

Table 1. The meaning of all notations in this paper

agents	explanations
Δ_{Ata}	agents responsible for Task Allocation.
Δ_{Adc}	Task Decomposition agents.
Δ_{Adv}	Task Division agents.
Δ_{Aa}	all the problem solving agents eligible for a task in the system.
Δ_{Aps}	problem solving agents for the unit-tasks.
Δ_{Aps}^{m}	problem solving agents for the multi-allocated unit-tasks.
Δ_{Aps}^{u}	problem solving agents for the unique-allocated unit-tasks.
Δ_{Ass}	Solution Synthesis agents.
Δ_{Ascm}	Solution Composition agents.
Δ_{Ascn}	Solution Construction agents.
tasks	**explanations**
Δ_{s-task}	sub-tasks resulting from the Task Decomposition for a task.
Δ_{s-task}^{m}	multi-allocated sub-tasks.
Δ_{s-task}^{u}	unique-allocated sub-tasks.
Δ_{p-task}	partial-tasks of a task.
Δ_{p-task}^{m}	multi-allocated partial-tasks.
Δ_{p-task}^{u}	unique-allocated partial-tasks.
solutions	**explanations**
Δ_{s-solu}	sub-solutions for the sub-tasks.
Δ_{s-solu}^{m}	sub-solutions for the sub-tasks that are multi-allocated.
Δ_{s-solu}^{u}	sub-solutions for the sub-tasks that are unique-allocated.
Δ_{p-solu}	partial-solutions for the partial-tasks.
Δ_{p-solu}^{m}	partial-solutions for the partial-tasks that are multi-allocated.
Δ_{p-solu}^{u}	partial-solutions for the partial-tasks that are unique-allocated.
Δ_{f-solu}	feature-solutions for the same unit-task.

2.2 A Framework for Task Unique-Allocation Problems (DPS1)

As described in Definition 2, Task Unique-Allocation is a process in which a task is allocated to a single agent.

The framework for task unique-allocation problems (DPS1) includes two phases: Task Unique-Allocation and Problem Solving. The procedure is shown in Fig. 1. In this framework, communication only exists between the task owner and the problem solving agent when the task owner first sends the task to the problem solving agent, and when the problem solving agent then sends the solution back to the owner.

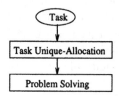

Fig. 1. The framework for task unique-allocation problems

Formal definitions using the Backus-Naur Form (BNF) specification and the formal expression are shown in Table 2.

Table 2. The framework for DPS1

BNF specification	corresponding formal expression
< **DPS1** > ::= < **TAu** >< **PS** >	
< TAu > ::= Task Unique-allocation	TAu : $\Delta_{Ata} \times \{\Delta_{Aa}, \{T\}\} \mapsto \{Aps\}$
< PS > ::= /* described in Table 6 */	PS : $\{Aps\} \times \{T\} \mapsto \{R\}$

2.3 A Framework for Task Multi-Allocation Problems (DPS2) and Solution Synthesis

In a task multi-allocation process, two or more agents are allocated to the same task (recall Definition 3) in order to increase reliability by avoiding bias. The framework for task multi-allocation problems (DPS2) includes three phases: Task Multi-Allocation, Problem Solving, and Solution Synthesis. In some cases, Problem Solving and Solution Synthesis can be iterated. The DPS2 framework is shown in Fig. 2 and Table 3.

Examples of multi-allocation problems include distributed medical expert systems [Zhang 92] in which a problem patient could be diagnosed by more than one expert system, and the collaborative Engineering Design domain [Bond 89]

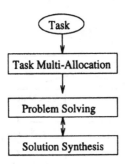

Fig. 2. The framework for task multi-allocation problems

Table 3. The framework for DPS2

BNF specification	corresponding formal expression
< DPS2 > ::= < TAm > < PS > < SS >	
< TAm > ::= Task Multi-Allocation	TAm : $\Delta_{Ata} \times \{\Delta_{Aa}, \{T\}\} \mapsto \Delta_{Aps}$
< PS > ::= /* described in Table 6 */	PS : $\Delta_{Aps} \times \{T\} \mapsto \Delta_{f-solu}$
< SS > ::= Solution Synthesis	SS : $\Delta_{Ass} \times \Delta_{f-solu} \mapsto \{R\}$

in which the cooperative experts have different areas of specialisation (i.e., one is expert in functionality, and the other in production).

Because these agents have different problem solving expertise, their solutions to this same task could have different features. Because some f-solutions may be consistent while some may be not, a Solution Synthesis process is expected to deal with different relationships among these f-solutions. The goal of SS is to gather the strength of every agent's solution and resolve conflicts when necessary. After this process reduces some bias and ambiguity, the final synthesised solution will be more reliable and precise provided that the SS process proceeds smoothly.

There are various different solution synthesis strategies. In [Brandau & Weihmayer 89], kitchen design is decided cooperatively by the architect and the cook. They may have competitive standpoints toward certain features of the kitchen. Negotiation is used to resolve their disagreements.

In a Distributed Expert System [Zhang 92], the inconsistency of beliefs that agents have for each possible solution can be resolved by using a mathematical model [Zhang & Zhang 94] or a neural network model [Zhang & Zhang 95]

2.4 A Framework for Task Decomposition Problems (DPS3) and Solution Composition

Task Decomposition process is defined in Definition 7. The framework for task decomposition problems (DPS3) includes four phases: Task Decomposition, Task-Allocation, sub-task Problem Solving, and Solution Composition. In some cases, Problem Solving and Solution Composition can be iterated. This is illustrated in Fig. 3(a).

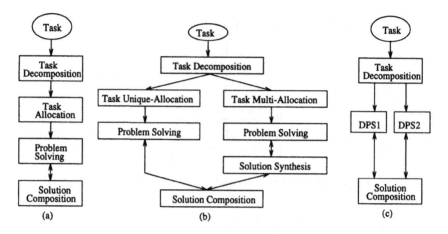

Fig. 3. The framework for task decomposition problems

Let us analyse DPS3 further. As we have mentioned in Section 2.2 and 2.3, Task Allocation has two styles: unique-allocation and multi-allocation. If a sub-task is allocated to two or more agents, then its sub-solution must be first obtained through Solution Synthesis process on the feature-solutions offered by these agents working on this sub-task. So the framework for DPS3 may be changed to Fig. 3(b). In fact, if a sub-task is multi-allocated, then the problem solving for this task takes the form of DPS2, and if a sub-task is unique-allocated, its problem solving takes the form of DPS1. So Fig. 3(b) can be abstracted to Fig. 3(c) using DPS1 and DPS2.

Generally, the framework for DPS3 is shown in Table 4.

Table 4. The framework for DPS3

BNF specification	corresponding formal expression
< **DPS3** > ::= < **TDc** > (< **DPS1** > | < **DPS2** >) < **SCm** >	
< TDc > ::= Task Decomposition	$TDc : \Delta_{Adc} \times \{T\} \mapsto \Delta_{s-task}$
	$\Delta_{s-task} = \Delta^u_{s-task} \cup \Delta^m_{s-task}$
< DPS1 > ::= /* described in Table 2 */	$TAu : \Delta_{Ata} \times \{\Delta_{Aa}, \Delta^u_{s-task}\} \mapsto \Delta^u_{Aps}$
	$PS : \Delta^u_{Aps} \times \Delta^u_{s-task} \mapsto \Delta^u_{s-solu}$
< DPS2 > ::= /* described in Table 3 */	$TAm : \Delta_{Ata} \times \{\Delta_{Aa}, \Delta^m_{s-task}\} \mapsto \Delta^m_{Aps}$
	$PS : \Delta^m_{Aps} \times \Delta^m_{s-task} \mapsto \Delta^m_{f-solu}$
	$SS : \Delta_{Ass} \times \Delta^m_{f-solu} \mapsto \Delta^m_{s-solu}$
< SCm > ::= Solution Composition	$\Delta_{s-solu} = \Delta^u_{s-solu} \cup \Delta^m_{s-solu}$
	$Scm : \Delta_{Ascm} \times \Delta_{s-solu} \mapsto \{R\}$

From the comparison of Fig. 3(b), (c), and Table 4, we can observe one important point: Problem Solving process in DPS3 can be DPS1 or DPS2, i.e., the

Problem Solving for sub-tasks can itself be a Distributed Problem Solving process. This is called the *recursive feature* of Distributed Problem Solving which is explicitly expressed in the comprehensive framework for DPS in Section 3.

An example of DPS3 is medical diagnosis. The task of diagnosing a patient who may have a liver disease is decomposed by the doctor into: ST_1, blood test; ST_2, blood pressure; and ST_3, body temperature. To carry out these three sub-tasks, the responsible experts should have different areas of expertise because these sub-tasks are of different types. Though one expert may be proficient in one type of task, he/she may not be capable of another type of task. These sub-solutions are then composed by the doctor in charge who makes a decision using his/her specialised expertise.

2.5 A Framework for Task Division Problems (DPS4) and Solution Construction

Task Division is defined in Definition 10. The framework for task division problems (DPS4) includes four phases: Task Division, Task-Allocation, partial-task Problem Solving, and Solution Construction. In some cases, Problem Solving and Solution Construction can be iterated. This is illustrated in Fig. 4 and Table 5.

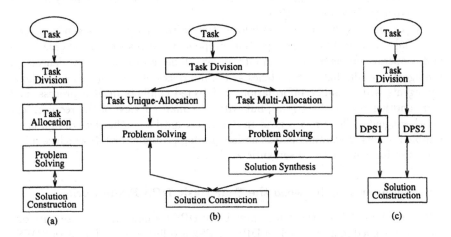

Fig. 4. The framework for task division problems

There are two main reasons why a task may be divided into several partial-tasks.

(1) The geographical feature of the task. In DVMT [Lesser & Corkill 83], agents sitting in different regions can only sense local data which is only part of the data set for the entire task. So the task is naturally divided into partial-tasks according to the sensing scope of agents. This is referred to as data-incomplete DPS4. Other examples are air traffic control [Steeb el al. 88], and traffic light control [Lesser et al. 79].

Table 5. The framework for DPS4

BNF specification	corresponding formal expression
< DPS4 > ::= < TDv > (< DPS1 >	
| < DPS2 >) < SCn >	
< TDv > ::= Task Division	TDv: $\Delta_{Adv} \times \{T\} \mapsto \Delta_{p-task}$
	$\Delta_{p-task} = \Delta^u_{p-task} \cup \Delta^m_{p-task}$
< DPS1 > ::= /* described in Table 2 */	TAu: $\Delta_{Ata} \times \{\Delta_{Aa}, \Delta^u_{p-task}\} \mapsto \Delta^u_{Aps}$
	PS: $\Delta^u_{Aps} \times \Delta^u_{p-task} \mapsto \Delta^u_{p-solu}$
< DPS2 > ::= /* described in Table 3*/	TAm: $\Delta_{Ata} \times \{\Delta_{Aa}, \Delta^m_{p-task}\} \mapsto \Delta^m_{Aps}$
	PS : $\Delta^m_{Aps} \times \Delta^m_{p-task} \mapsto \Delta^m_{f-solu}$
	SS : $\Delta_{Ass} \times \Delta^m_{f-solu} \mapsto \Delta^m_{p-solu}$
< SCn > ::= Solution Construction	$\Delta_{p-solu} = \Delta^u_{p-solu} \cup \Delta^m_{p-solu}$
	SCn : $\Delta_{Ascn} \times \Delta_{p-solu} \mapsto \{R\}$

PGP [Durfee & Lesser 87, 89] is used as an efficient approach to result sharing and solution construction in DVMT. When nodes have the same global goal but different PGPs, they will combine these PGPs into a compromised PGP. When they fail to reach an agreement, nodes with high authority will have the responsibility to decide it.

(2) The agents have only partial expertise for a whole task. This is referred to as expertise-incomplete DPS4.

An example is the distributed information retrieval system [Nagendraprasad et al. 95] where agents retrieve partial component designs from local case bases and cooperatively assemble them into an overall design for steam condensers. If the agents' results have mismatches at the interfaces of the component designs, the agents will exchange information on detected conflicts in SCn process, and start another round of retrieval, but with an enhanced view of the requirement of other agents. This process goes on iteratively until a conflict free design is assembled. Another example is text understanding [Wang et al. 94] which requires expertise for syntactic and semantic analysis.

2.6 Comparisons between the Four Basic DPS Frameworks

DPS1 is different from the other three basic DPS frameworks because only one agent is involved in a unit-task in DPS1. DPS2 is different from DPS3 and DPS4 in that agents in DPS2 all have complete expertise for the problem solving of a unit-task, and they all solve the whole unit-task independently; but agents in DPS3 and DPS4 only deal with certain parts of a unit-task either because of the incomplete task data they are able to access to, or incomplete expertise they have for the problem solving. The differences between DPS3 and DPS4 are as follows.

In DPS3, the sub-solutions in decomposition problems together can not be naturally formed into the solution for the original task. In other words, Solution Composition is always necessary for the task to be completed. Generally, it is a high-level decision making process which requires the sub-solutions as its supporting data (e.g., in the medical diagnosis example, the doctor uses the available data from the sub-solutions to make a diagnosis decision applying his own diagnosis expertise). Secondly, Solution Composition generally take place after all the sub-solutions are available.

However, in DPS4, Solution Construction is applied to partial-solutions in order for them to meet the *domain* or *consistency constraints* [Erman et al. 80]. All partial-tasks are in the same abstract level. SCn can take place after or during the partial-task problem solving when some responsible agents find their partial solutions are inconsistent or violate some domain constraints. Moreover, in some division problems where there are no such constraints between partial-tasks (e.g., traffic light control [Lesser et al. 79], or all the partial-solutions have already met all the constraints, SCn can even be not necessary; i.e., the partial-solutions are simply put together to form the global solution for the original task.

3   A Comprehensive Framework for DPS and Solution Integration

3.1   The Definition of a Comprehensive Framework for DPS

Based on Section 2, the characteristics of a comprehensive DPS framework can be summarised as:

(1) The task can be transferred from one agent to another: (a) without previously being decomposed or divided, or (b) after being decomposed, or (c) after being divided;

(2) The whole task, sub-tasks, or partial-tasks to be transferred can be assigned or allocated to: (a) one agent, or (b) more than one agent.

(3) The agents responsible for the transferred whole task, sub-tasks or partial-tasks begin their problem solving either through: (a) concentrating on their domain problem computing (DPC, see Definition 17 below); or, (b) recursively using the same cycle as (1) → (2) → (3) → (4), i.e., further transferring the whole current task, or further decomposing or dividing the current task into smaller parts and then allocating them to other agents.

(4) The solutions relevant to the task are integrated into a more comprehensive whole.

(a) If a task is transferred as a whole to only one agent, then the integration is not required. The original owner of the task only has to pick up the solution from the allocated agent.

(b) Otherwise, the relevant solutions are: all feature-solutions if the task is multi-allocated; or all sub-solutions if the task is decomposed; or all partial-solutions if the task is divided.

[Definition 17] Domain Problem Computing (DPC) is a process that the problem solving agent actually solves its task concerned with only the problem domain.

The comprehensive framework for DPS is illustrated in Fig. 5 and Table 6.

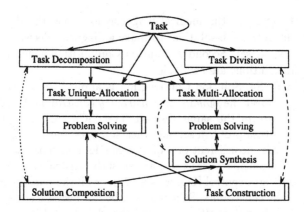

Fig. 5. The comprehensive framework for DPS

Table 6. The BNF specification of a comprehensive DPS framework

```
< DPS > ::= < DPS1 > | < DPS2 > | < DPS3 > | < DPS4 >
< DPS1 > ::= < TAu >< PS >
< TAu >   ::= Task Unique-Allocation
< PS >    ::= < DPC > | < DPS >
< DPC >   ::= Domain Problem Computing
< DPS2 > ::= < TAm >< PS >< SS >
<TAm >    ::= Task Multi-Allocation
< SS >    ::= Solution Synthesis
< DPS3 > ::= < TDc > (< DPS1 > | < DPS2 >) < SCm >
< TDc >   ::= Task Decomposition
< SCm >   ::= Solution Composition
< DPS4 > ::= < TDv > (< DPS1 > | < DPS2 >) < SCn >
< TDv >   ::= Task Division
< SCn >   ::= Solution Construction
```

3.2 Solution Integration

Distribution, incompleteness and diversity of knowledge, data, and functionality of problem solving agents result in agents' incomplete, inconsistent or even contradictory solutions to a task. Solution Integration (SI) is an important process that deals with the above issues when it combines integral solutions to achieve a comprehensive solution to the task. When integrating f-solutions, SI is featured as SS in multi-allocation problems; when integrating s-solutions, SI is refereed to as SCm in decomposition problems; and when integrating p-solutions, SI is distinguished as SCn in division problems.

There are many factors affecting SI on its strategies, processing data, processing time, and time schedule, especially that imposed by domain constraints, task interdependence, agent authorities, and user constraints.

(1) Domain Constraints: Domain Constraints decide the domain features of Solution Integration. For example, in some domains, Solution Integration needs

little computation (e.g., traffic light control in [Lesser et al. 79]); while in other domain, Solution Integration may need significant computation and communication to satisfy domain constraints (e.g. in HEARSAY-II [Erman et al. 80]).

(2) Task Interdependency: For partial- (or sub-) tasks $PT_i(ST_i)$ and $PT_j(ST_j)$, (a) if after $PT_i(ST_i)$ is solved, it would be not necessary to solve $PT_j(ST_j)$, then we say that they have logical constraints; and (b) if $PT_j(ST_j)$ can only start its problem solving after $PT_i(ST_i)$ has been solved, then we say that they have time order constraints. One fairly comprehensive formalisation of task interdependencies can be found in TARMS [Decker & Lesser 93] where the task interdependencies are referred to as non-local effects (e.g. *enables, facilitates, hinders* and *precedence* in modelling different task environments).

(3) Agent Authorities: When multiple agents interact with each other in the process of cooperative group decision making, there may exist interdependencies (e.g. different authorities) among agents. Consider a medical diagnosis example. Let A be a doctor in charge of a case, and B be an assistant. The decision of B may be constrained by $A's$ decision.

(4) User Constraints: There are two user constraints which could be provided by a user on the final solution: time constraints (when a solution can be obtained) and quality constraints (for details, see [Yang & Zhang 95]). Generally, Solution Integration begins after the integral results are all available. But in a real system, the system must allow different favourite time limits users give. If a user gives an earlier deadline, Solution Integration agent(s) must arrange its (their) own timetable even though integral results are incomplete. Otherwise, the user will get nothing when the deadline comes.

4 Conclusion

In order to clarify different types of DPS frameworks, this paper has defined four basic types of DPS frameworks, and a comprehensive DPS framework accordingly. This work included the framework proposed by Smith and Davis [Smith 80; Smith & Davis 81; Davis & Smith 83] DPS3 for task decomposition problems. DVMT [Lesser & Corkill 81, 83; Durfee & Lesser 87, 89], HEARSAY-II [Erman et al. 80]; and traffic control systems [Lesser et al. 79; Steeb et al. 88], etc., are classified as instances of DPS4 for task division problems. Cooperative experts working on a same task [Brandau & Weihmayer 89; Zhang 92; Wang et al. 94] are examples of DPS2 for task multi-allocation problems. DPS1, the simplest and most basic framework for task unique-allocation problems, can exist in traditional knowledge-based systems and other DPS frameworks. So the comprehensive framework can cover single and multiple agent systems.

As the second contribution of this paper, three different types of Solution Integration have also been clarified. We also examined important factors that affect SI strategies, processing data, and processing time. We believe that this part of the work is also very important, as previous papers only concentrated on strategies in a particular context.

Just as one of DPS techniques is about how autonomous agents interact with each other in a DPS system, our next attempt is to examine how different, "autonomous" DPS systems (with possible different frameworks) interact in multi-agent systems.

References

[Bond 89] Bond A. H., "The Cooperation of Experts in Engineering Design", Distributed Artificial Intelligence, Vol. 2, Gasser L. and Huhns M. N., eds., Pitman Publishing, pp. 463-486, 1989.

[Brandau & Weihmayer 89] Brandau R. and Weihmayer R., "Heterogeneous Multiagent Cooperative Problem Solving in a Telecommunication Network Management Domain", in Proc. of the 9th Workshop on Distributed Artificial Intelligence, 1989.

[Davis & Smith 83] Davis R. and Smith R., "Negotiation as a Metaphor for Distributed Problem Solving", Artificial Intelligence, Vol. 20, pp. 63-109, 1983.

[Decker & Lesser 93] Decker K. S. and Lesser V. R., "Quantitative Modeling of Complex Computational Task Environment", in Proc. of the 11th National Conference on Artificial Intelligence, pp. 217-224, 1993.

[Durfee & Lesser 87] Durfee E. H. and Lesser V. R., "Using Partial Global Plans to Coordinate Distributed Problem Solvers", in Proc. of the 10th International Joint Conference on Artificial Intelligence, pp. 875-883, 1987.

[Durfee & Lesser 89] Durfee E. H. and Lesser V. R., "Negotiating Task Decomposition and Allocating Using Partial Global Planning", Distributed Artificial Intelligence, Vol. 2, Gasser L. and Huhns M. N., eds., Pitman Publishing, pp. 229-244, 1989.

[Erman et al. 80] Erman L. D., Hayes-Roth F., Lesser V. R., and Reddy D. R., "The Hearsay-II Speech-Understanding System: Integrating Knowledge to Resolve Uncertainty", Computing Surveys, Vol. 12, pp. 213-253, June 1980.

[Hewitt 91] Hewitt C., "Open Information System Semantics for Distributed Artificial Intelligence", Artificial Intelligence, Vol. 47, pp. 79-106, 1991.

[Kornfeld & Hewitt 81] Kornfeld W. A. and Hewitt C., "The Scientific Community Metaphor", IEEE Trans. Syst., Man, Cybern., Vol. SMC-11(1), pp. 24-33, Jan. 1981.

[Lesser et al. 79] Lesser V.R., et al., "Working Papers in Distributed Computation 1: Cooperative Distributed Problem Solving", Dept. Compt. and Inform. Sci., Univ. of Massachusetts, Amherst, July 1979.

[Lesser & Corkill 81] Lesser V. R. and Corkill D. D., "Functionally Accurate Cooperative Distributed Systems", IEEE Trans. Syst., Man, Cybern., Vol. SMC-11, pp. 81-96, 1981.

[Lesser & Corkill 83] Lesser V. R. and Corkill D. D., "The Distributed Vehicle Monitoring Testbed: A Tool for Investigating Distributed Problem Solving Networks", AI Magazine, Vol. 4, pp. 15-33, 1983.

[Lesser 91] Lesser V. R., "A Retrospective View of FA/C Distributed Problem Solving", IEEE Trans. Syst., Man, Cybern., Vol. SMC-21(6), pp. 1347-1362, 1991.

[Nagendraprasad et al. 95] Nagendraprasad M. V., Lesser V. R., and Lander S., "Reasoning and Retrieval in Distributed Case Bases", UMASS Technical Report 95-27, 1995.

[Smith 80] Smith R. G., "The Contract Net Protocol: High-level Communication and Control in a Distributed Problem Solver", IEEE Trans. Comput., Vol. C-29, pp. 1104-1113, 1980.

[Smith & Davis 81] Smith R. G. and Davis R., "Framework for Cooperation in Distributed Problem Solving", IEEE Trans. Syst., Man, Cybern., Vol. SMC-11, pp. 61-70, 1981.

[Steeb et al. 88] Steeb R., Cammarata S., Hayes-Roth F., Thorndyke P., and Wesson R., "Architecture for Distributed Intelligence for Air Fleet Control", Readings in Distributed Artificial Intelligence, Bond and Gasser, eds., Morgan Kanfmann Publisher, San Mateo, CA, pp. 90-101, 1988.

[Wang et al. 94] Wang K., Hu P., Yang H., Kang X., et al., "Text Understanding-oriented, Multi-agent System MAS/TH-3", 863-Project Report, Dept. of Computer Science and Technology, Tsinghua University, April 1994.

[Yang et al. 85] Yang J-Y D., Huhns M. N., and Stephens L. M., "An Architecture for Control and Communications in Distributed Artificial Intelligence Systems", IEEE Trans. Syst., Man, Cybern., Vol. SMC-15(3), pp. 316-326, 1985.

[Yang et al. 93] Yang, H., Shi C., Hu P., and Wang K., "A Method of Solution Synthesis in Open Distributed System ODKPS/TH-3", in Proc. of Artificial Intelligence'93, pp. 71-77, Nanjing, China, 1993.

[Yang & Zhang 95] Yang H. and Zhang C., "Satisfying User Constraints on Solution in Multi-agent Systems", in Proc. of the 8th Australian Joint Conference on Artificial Intelligence, World Scientific Publisher, pp. 187-194, Canberra, 1995.

[Zhang 92] Zhang C., "Cooperation under Uncertainty in Distributed Expert Systems", Artificial Intelligence, Vol. 56, pp. 21-69, 1992.

[Zhang & Zhang 94] Zhang M. and Zhang C., "A Comprehensive Synthesis Strategy for Conflict Resolutions in Distributed Expert Systems", Australian Journal of Intelligent Information Processing Systems, Vol. 1, No. 2, pp. 21-28, 1994.

[Zhang & Zhang 95] Zhang M. and Zhang C., "A Neural Network Strategy for Solving Belief Conflict in Distributed Expert Systems", in Proc. of 4th International Conference for Young Computer Scientists, pp. 429-436, 1995.

A Logical Framework for Multi-Agent Systems and Joint Attitudes*

Lawrence Cavedon[†] & Gil Tidhar[‡]

[†]Dept. of Computer Science, R.M.I.T.
GPO Box 2476V, Melbourne, Australia
cavedon@cs.rmit.edu.au

[‡]Australian Artificial Intelligence Institute
171 LaTrobe St., Melbourne, Australia
gil@aaii.oz.au

Abstract. We present a logical framework for reasoning about multi-agent systems. This framework uses Giunchiglia et al.'s notion of a logical *context* to define a methodology for the modular specification of agents and systems of agents. In particular, the suggested methodology possesses important features from the paradigm of object-oriented (OO) design. We are particularly interested in the specification of agent behaviours via BDI theories—i.e., theories of *belief, desire* and *intention*. We explore various issues arising from the BDI specification of systems of agents and illustrate how our framework can be used to specify bottom-level agent behaviour via the specification of top-level intentions, or to reason about complex "emergent behaviour" by specifying the relationship between simple interacting agents.

1 Introduction

The formal specification of autonomous reasoning agents has recently received much attention in the AI community, particular under the paradigm of *agent-oriented programming* [3]. In this paper, we are specifically concerned with issues that arise in the formal specification of multi-agent systems, particularly *teams* of agents that are designed to exhibit specified behaviour as a team. The framework that we describe addresses issues that arise when teams may consist of truly *heterogeneous* agents—i.e., agents which may view the world differently to each other and which may exhibit different characteristics. Logical systems for such systems of agents have tended to have a "global" nature—a single logical framework (involving a single logical language) is used and each agent's behaviour is

---

* The authors wish to thank David Morley, Anand Rao, Liz Sonenberg, Lin Padgham, and Mike Georgeff for their valuable contributions and suggestions in the discussions that led to this paper. An anonymous referee also made several useful suggestions. This work was partially supported by the *Cooperative Research Centre for Intelligent Decision Systems*, Melbourne, Australia.

specified within that framework. We propose the use of a *contextualised* logical framework—i.e., a *logic of context*—to formally model systems of agents.

Various authors (e.g., Guha [9] and McCarthy [15]) have recently argued that accounting for *context* is crucial when defining models of efficient reasoning. We apply this idea to the multi-agent scenario by ensuring each agent is modelled by a separate context, and allowing each such context to involve its own logical language. This promises a number of potential advantages when designing agents:

- Each agent can be made as simple as possible: in particular, an agent need only be concerned with the propositions in the world that are directly related to its own reasoning;
- Different agents can be modelled as having different discriminatory abilities, or at least different perspectives on the world;[2]
- Each agent can be made more computationally efficient by limiting the scope of its reasoning;
- By providing means for succinctly specifying the interaction between agents, a more modular approach to the design of multi-agent systems is obtained.

Morley et al. [16] have addressed several of the above concerns within a partial logic framework based on *situation semantics*—each agent need only be concerned with partial aspects of the world and is completely agnostic with respect to other conditions and events happening in it. We consider our approach to be a natural extension of theirs—different agents are effectively modelled using different languages rather than by partitioning a single language across different agents. Our framework also provides a more rigid discipline for distributed design and specification since the interaction between the logical contexts representing agents can be constrained in ways that promote a strict hierarchical design approach—i.e., a formal model for each agent is separately defined, followed by the definition of the interaction between them.

The logic of context we use here is that of Giunchiglia et al. [5, 7]. This framework is particularly flexible in that different contexts can involve different logical languages, which is particularly relevant to our enterprise. We use the framework as the basis for the specification of multi-agent systems in which any agent can be logically specified independently of the others. In fact, we generalize this notion to allow a complex autonomous system to be hierarchically specified in terms of simpler component systems. *Bridge rules*—rules of inference which define the interaction between contexts—provide a mechanism for specifying the interaction between independent agents/components. This framework for modular modelling and specification suggests a methodology for the distributed design of systems of agents that corresponds to the design principles of object

---

[2] This parallels recent work in modelling human dialogue in which it is claimed that a model of interacting systems must provide different agents with the ability to discriminate the world differently [10]. Of related interest from the AI literature is work by Lesperance and Levesque [12] who define a logical model of *indexical* knowledge and reasoning.

oriented systems (c.f. Booch [2]) and is an important motivating factor in our choice of framework.

One use of the proposed framework that we are particularly interested in is in logically specifying agents using a BDI (belief-desire-intention) architecture, following Cohen and Levesque [6] and Rao and Georgeff [17, 18]. If each context in a system corresponds to a BDI theory, then bridge rules can be used to specify how the beliefs, desires and intentions at the agent level relate to the beliefs, desires and intentions at the team level. In particular, our framework can incorporate either a *reductionist* model of behaviour (whereby the behaviour of a team or system can be completely explained by the behaviour of the component agents—e.g., Grosz and Sidner [8]) or a *holistic* model (whereby the behaviour of a system cannot be explained solely in terms of its components—e.g., Searle [20]). We discuss these issues below and show how we can naturally model both top-down specification and bottom-up emergent behaviour.

2 Logics of Context

The role of *context* in representation and reasoning has for some time been seen as central to efficiency in the linguistic and philosophical literature (e.g., see Barwise [1]). The importance of context in reasoning has also assumed prominence in AI, with a number of different "logics of context" appearing in the recent literature [7, 9, 15]. These logics have mainly been used to define modular models of logical reasoning, leading to more efficient inferencing. The logical framework developed by Giunchiglia and his co-workers [7, 5] is particularly flexible. In their framework, a context is more than a collection of facts—a context is a logical theory, with its own (private) language. This flexibility is precisely what is needed for our purposes: by letting each agent be modelled by its own context, we allow different agents to be reasoning about completely different aspects of the world, with no overlap (if that is what is desired).

The definition of Giunchiglia et al.'s framework is as follows. Note that their framework is purely syntactic. A rigorous model-theoretic semantics is an important topic for future research, which will also address some of the problems raised below.[3]

Definition 1. (Giunchiglia et al.) A *context* is a logical theory, represented as a triple $\langle L, Ax, \Delta \rangle$, where L is a logical language, Ax is a set of axioms over L and Δ is a set of inference rules. An inference rule will be written as $\frac{F_1}{F_2}$ where F_1 (the antecedent) and F_2 (the consequent) are formulas of L. □

As can be seen, each context involves its own language. The logical theory of a context is specified by defining its axioms and the rules of inference under which

---

[3] Giunchiglia et al. are also currently working on a model-theoretic semantics for their framework (private communication).

it is closed. Giunchiglia et al. explicitly allow each different context to contain different inference rules—in fact, one context may contain inference rules for a classical logic while another context's inference rules are for default reasoning, or some other non-classical system. This approach provides great flexibility and we are certainly happy to support it. However, for the purposes of the current paper, we will generally assume that the inference rules of classical logic (or modal logic where appropriate) are the only ones contained in a context.

Information in one context can constrain or relate to information in another context. Giunchiglia et al. introduce the notion of *bridge rules* which are used to build informational bridges across contexts. In general, a bridge rule is written as follows:

$$\frac{C:F}{C':F'}$$

This rule ensures that if F holds in context C, then F' also holds in context C'. Bridge rules are actually "meta-contextual" entities, in that a bridge rule is not contained in any particular context. However, we will assume that any such rule is associated with some context C''. In fact, the definition of a *multi-system* below ensures that our use of bridge rules is reasonably strictly regulated in this way. The following sorts of bridge rules are particularly important:

$$\text{lifting rule: } \frac{C:F}{F''} \qquad \text{lowering rule: } \frac{F''}{C:F}$$

The context involved in the consequent (resp., antecedent) of a lifting (resp., lowering) rule is taken to be the current context—i.e., the context C'' with which the rule is associated. A lifting rule allows information embedded in a sub-context of C'' to be "lifted" into C''—i.e., to ensure that it is reflected in some way in C''; a lowering rule achieves the converse—i.e., it ensures that certain information in C'' is reflected in some sub-context of it. Bridge rules provide the interface between contexts and therefore between agents and between systems of agents. As such, they play an important role in the model defined below.

Cimatti and Serafini [5] outline a proof technique for "situated reasoning" based on the logic of context framework. A formula p is "lowered" into a context C, consequences of p may be derived *within* C, and (some of) these consequences are then lifted back into the original context. However, the reasoning performed in C may be kept completely private—all that needs to be made known to the higher context is the theorems that are required. For example, suppose the outer context O has a lowering rule $\frac{r}{C:p}$, where C is a component context. Further suppose that s follows from p in C and that t follows from s in C. Then $p \supset t$ is a theorem of C. Now suppose that $\frac{C:t}{q}$ is a lifting rule in O. It follows that $C:p \vdash C:t$ and therefore that $r \vdash q$ in O. However, the reasoning in C which led to this consequence is hidden from context O—in particular, the proposition s is not in any way present in the logical language of O. In terms of our multi-agent model, the agent corresponding to O needs only to know about the most abstract issues concerning the C-agent (in this case, that C will infer t from p) without *any* knowledge of the internal workings of the C-agent.[4]

---

[4] See Cimatti and Serafini [5] for further details of this contextualised proof technique.

3 Reasoning About Events

In the initial version of our framework, the interaction between agents is restricted to the causation of events. For example, an event occurring in one agent's context may trigger an event in some other agent's context. This restriction ensures maximal encapsulation of each agents' reasoning and knowledge about the world. Any such interaction is, of course, modelled using bridge rules. For example, there may be a bridge rule stating that a *dial-marys-number* event in John's context causes a *telephone-rings* event in Mary's context.

Our motivation for such a view is recent work by Morley et al. [16], who define a framework for reasoning about events based on *situation semantics* (e.g., see Barwise [1]). Morley et al.'s concern is to model the multitude of events that may be taking place at any given point in time. Any particular agent, however, need only be concerned with events and conditions which are "local" to it. In particular, this allows the specification of simple agents—since an agent does not need to be able to reason about events and states of the world that do not influence its behaviour, these can be left out of its model of the world.

Morley et al. use a partial logic to model this stance—any agent may be "ignorant" with respect to some aspects of its world state, in that there are propositions that are neither true nor false. Systems of such agents can be combined with different agents being ignorant about different world conditions. However, the logic is "global" (in the sense that there is one language in which all agents are modelled) so as to ensure that the agents can interact—if there was no overlap between the conditions with which agents were knowledgeable, then one agent's behaviour could not affect any other agent.

We use contexts to effectively model Morley et al.'s framework. Rather than have an agent be ignorant about some "global" world conditions and knowledgeable about others, we provide each agent with its own collection of conditions—i.e., its own logical language. Also, no agent's language need overlap with any others'—any interaction between agent behaviours is specified by bridge rules. This way of representing collections of agents is both flexible—each agent has its own language and can be reasoned about completely independently from the existence of all other agents—but also regimented—interaction is limited to the use of bridge rules, and by the specific restriction we defined above.

Since we model each agent's *limited awareness* by assigning it a context (as opposed to using a partial logic), we can make use of a standard logic for reasoning about events and actions, along the lines of those used by Lifschitz [13] and others. In particular, we assume a set of *fluents* C denoting properties that change over time, a set of *values* B of those fluents, a set $\mathcal{E}$ of *event types*, and a set $\mathcal{I}$ of *time points*. Associated with each event type $e \in \mathcal{E}$ are two sets of fluent-value pairs, the *precondition* and *postcondition* of e. For example, the precondition and postcondition of the event *close* of closing a door could be defined as follows:

event	*pre*	*post*
close	*closed* : F	*closed* : T

Morley et al. define a number of axioms which their logic for reasoning about events is required to satisfy. On top of the standard axioms of first-order logic, these are the following:

$$holds(c, b, i) \wedge holds(c, b', i) \supset b = b' \qquad \text{(Coherency)}$$
$$occurs(e, i, i') \wedge pre(e, c, b) \supset holds(c, b, i) \qquad \text{(Precondition Occurrence)}$$
$$occurs(e, i, i') \wedge post(e, c, b) \supset holds(c, b, i') \qquad \text{(Postcondition Occurrence)}$$
$$holds(c, b, i) \wedge \neg holds(c, b, i') \wedge i < i' \supset \qquad \text{(Observable Change)}$$
$$\exists b' \exists e \exists i_1 \exists i'_1 . i \leq i_1 < i'_1 \leq i' \wedge (occurs(e, i_1, i'_1) \wedge post(e, c, b'))$$

The last of these axioms is effectively an *axiom of persistence* stating that an agent will infer a change in some world-condition only if the agent detects an event that it knows affects that condition. This simple model of events is sufficient for our illustrative purposes.[5]

Finally, Morley et al. present a method for composing *histories* of agents into *systems* of interacting agents. A history is a description of how the universe changes over time (for each agent). To compose systems, the histories corresponding to each agent are merged in such a way as to provide a consistent description of a changing world across the system of agents—i.e., the histories of the agents must be mutually consistent. The concept of a history is not required in the context-logic framework—the interaction between agents is specified by bridge rules which model the causation of events between different contexts. However, there are some remaining temporal issues to be resolved in our framework, as is discussed below.

4 Specifying Systems of Agents

In this section, we use Giunchiglia et al.'s logic of context to specify *multi-agent systems*. Informally, a system of agents is logically represented as a collection of contexts, where each context contains a logical specification of that particular agent's behaviour. Each agent may be specified independently of all others; merging them as a system involves providing the bridge rules which define the interaction between agent behaviours. Since each context involves a logical specification, we are able to reason about the behaviour of each agent. Using the contextualised method of reasoning outlined in Section 2, we can also use the bridge rules to reason about the behaviour of the system as a whole as a (heterogeneous) logical theory. Recall that bridge rules are to be restricted to involve causation of events across contexts. This ensures modularity between systems—i.e., each agent need only reason about the events it is likely to encounter and if some other agent's behaviour affects that of the first agent, then this is specified by providing a link between relevant events in each agent's context.

---

[5] Providing a "scope" for the effects of events, which is effectively what is provided through the use of contexts, has recently been claimed to address certain problematic issues in reasoning about events and their effects [14, 16].

As well as allowing systems of agents to be defined in terms of the behaviours of individual agents, we also want to allow such a system to be a component in a more complex system. That is, we want to provide a framework of *hierarchical* specification, whereby an agent may be specified in terms of its components, with each component also being hierarchically specified, and so on. We believe this to be a very powerful technique for the specification and design of complex reasoning systems. We restrict the use of bridge rules so that they can only communicate between a single level of hierarchy—as with our previous restriction on bridge rules, this ensures a regimented modular design approach to interacting systems.

The following definition captures the above intuitive notions. Since we have generalized a multi-agent system to one where an agent is itself a combination of simpler components, we refer to the compound system as a *multi-system*.

Definition 2. A *multi-system* S is a tuple $\langle C, \Delta, S_1, ..., S_n \rangle$, $n \geq 0$, where C is a context (the *context of S*) specifying a particular logical theory, Δ is a set of bridge rules, and each S_i is itself a multi-system with associated context C_i. All lowering rules are of the form $\dfrac{occurs(e, i)}{C_k : occurs(e', i')}$, lifting rules of the form $\dfrac{C_k : occurs(e, i)}{occurs(e', i')}$, and other bridge rules of the form $\dfrac{C_k : occurs(e, i)}{C_l : occurs(e', i')}$, where C_k, C_l are the contexts of some S_k, S_l, $0 \leq k, l \leq n$. $\square$

A note needs to be made regarding the temporal arguments in the bridge rules of this definition. The language of each context is, as pointed out several times earlier, private to that language. Hence, the time variables in the premise and consequent of a bridge rule are also unrelated to each other. This raises issues regarding coordination and synchronization. If each temporal variable will be interpreted (model-theoretically) within its own context, then no relationship can be imposed between the different associated time-lines. In particular, this results in a model of completely asynchronous interaction—events occur in one context and a corresponding event occurs in another context, at some completely unrelated time.[6] This view of multi-agent interaction certainly has its appeal. However, we may wish to be able to coordinate between contexts—in particular, it is not clear how we quantify over the temporal variables, and whether we can enforce one-to-one correspondence of events.[7] For this reason, we believe that a *global* view of time will sometimes prove useful. This issue can only be properly resolved once a model-theoretic semantics is provided, and hence we ignore the technicalities for now—for the purposes of this paper, we simply assume that temporal variables are global.

---

[6] Of course, what is important is that events *within* a context can be coordinated.

[7] For example, if we existentially quantify over the temporal variable, then a single occurrence of an event in one context will satisfy the existence required for two events in another context.

The restrictions on bridge rules that we mentioned earlier are reflected in the condition imposed on them in the above definition. All interaction between component systems is specified via bridge rules. Note also that the components of the system S are hidden from view outside of S, i.e., hidden from any other multi-system of which S itself may be a component. However, a system $\mathcal{L}$ of which S is a component may still communicate with the components of S so long as it does so via an interface which is explicitly provided in S—i.e., $\mathcal{L}$ may lower an event into S which in turn lowers an event into one of its components.

The above definition captures a number of important design features which have been borrowed from the object-oriented paradigm [2], namely: *hierarchy* (the system is specified in terms of its components); *encapsulation* (the reasoning performed in each component system is private to that system); *modularity* (the interface between component systems is clearly defined by the bridge rules); and *abstraction* (only (certain) results of reasoning processes are visible to higher-level systems).

5 A Simple Example

The hierarchical methodology for specification inherent in the definition of the previous section is best illustrated via the use of a simple example. Each system in the example involves a simple theory for reasoning about events, as described above in Section 3.

The **Automatic Teller Machine** (ATM) involves three systems: Mary (q_m), a Bank (q_b), and an ATM (q_a). Each of these systems has its own language, axioms and reasoning rules that describe it. Mary has an ATM card and would like to get some money from the ATM. Mary has the fluents *card-in* and *money* and events that are described in Table 1 (left).[8] The Bank controls cash and determines if there is cash available for withdrawal or not. It has only one fluent, *cash*, and events that are described in Table 1 (right).

Mary			Bank		
event	pre	post	event	pre	post
insert	card-in : F	card-in : T	*check*		
remove	card-in : T	card-in : F	*deposit*		cash : T
key	card-in : T		*reject*	cash : F	
get		money : T	*withdraw*	cash : T	cash : F

Table 1. Events for the system Mary (left) and the system Bank (right)

---

[8] Note that Mary needs an internal representation of the card being in or out only if she needs to reason about it. Omitting the card from the system will not affect the interaction between Mary and the ATM since she will still perform the events which interface with the ATM.

The ATM is a more complex system. It interacts with Mary and the Bank and is actually responsible for dispensing cash. It has a number of fluents and events that describe its behaviour. The fluents of the ATM are *card-in*, *pin*, *amount*, *window-open*, *card-valid*, and *cash-in-tray* and its events are those described in Table 2.

event	pre	post
pull		*card-in* : T
push		*card-in* : F
open	*card-in* : T	*window-open* : T
close	*card-in* : F	*window-open* : F
type	*window-open* : T	*pin* : T, *amount* : T
validate	*card-in* : T, *pin* : T	*card-valid* : T
check	*amount* : T	
count		*cash-in-tray* : T
dispense	*cash-in-tray* : T	*cash-in-tray* : F

Table 2. Events for the system ATM

The above three systems can each be viewed as a closed system with its own language and rules governing its behaviour, allowing the specification of each system to be completely independent of the specification of the others.

The interaction between the individual systems is specified by the definition of a number of bridge rules, resulting in the multi-system q_c (Cash) $\equiv \langle C_{q_c}, \Delta_{q_c}, q_m, q_b, q_a \rangle$. The language of the context C_{q_c} contains the event *transfer* and no fluents. Δ_{q_c} contains the following bridge rules:

1. $\dfrac{occurs(transfer, i)}{C_{q_b} : occurs(deposit, i')}$

2. $\dfrac{C_{q_m} : occurs(insert, i)}{C_{q_a} : occurs(pull, i')}$

3. $\dfrac{C_{q_a} : occurs(push, i)}{C_{q_m} : occurs(remove, i')}$

4. $\dfrac{C_{q_m} : occurs(key, i)}{C_{q_a} : occurs(type, i')}$

5. $\dfrac{C_{q_a} : occurs(check, i)}{C_{q_b} : occurs(check, i')}$

6. $\dfrac{C_{q_b} : occurs(reject, i)}{C_{q_a} : occurs(push, i')}$

7. $\dfrac{C_{q_b} : occurs(withdraw, i)}{C_{q_a} : occurs(count, i')}$

8. $\dfrac{C_{q_a} : occurs(dispense, i)}{C_{q_m} : occurs(get, i')}$

These bridge rules specify the interaction between the subsystems. For example, Mary and the ATM interact via the events *insert*, *remove*, *key*, and *get*. Thus the event *get* in Mary depends on the event *dispense* in the ATM which in turn depends on the event *withdraw* in the Bank. If *cash* does not hold in the Bank then Mary will remain in a state where *money* does not hold. The language, axioms, and rules that are part of C_{q_c} provide the mutual beliefs, joint goals, and joint intentions of this multi-agent system.

Now consider an employer of Mary as another system q_e (Employer) that includes the event *payment* (and possibly other events and fluents). Consider also a further new multi-system q_f (Finance) $\equiv \langle C_{q_f}, \Delta_{q_f}, q_c, q_e \rangle$, and suppose that Δ_{q_f} contains the bridge rule $\dfrac{C_{q_e} : occurs(payment, i)}{C_{q_c} : occurs(transfer, i')}$. In this system, a *payment* event in the Employer system will affect the state of the Bank which will in turn affect the state of Mary. This can be deduced by using the bridge rules and reasoning in different systems. It seems then that we have found a way for Mary to have money—she needs an employer.[9]

6 BDI Architectures and Team Behaviour

Each context in a system is a logical specification of an agent. So far, such a logical theory has been assumed to be a simple theory for reasoning about events, as described in Section 3. In this section, we discuss the viability of taking the theories to be logical specifications of behaviour in terms of the attitudes *belief, desire* and *intention* (BDI) [6, 17, 18]. The BDI specification of an agent constrains the way it acts in situations it may encounter, given a background library of plans. Informally, *beliefs* specify what the agent knows about the world and what it thinks is possible; *desires* specify what states of the world the agent would like to see attained in the future; while *intentions* relate to the goals of the agent itself, i.e., those particular future world states which the agent commits itself to bringing about. BDI logics are based on multi-modal logics with possible-worlds semantics. In our (syntactic) framework, we will simply use the modalities of belief (BEL) and intention (INT) without providing a model-theoretic semantics for them.[10]

When we move to the multiple agent case, more complex issues are raised regarding collaborative, or *team*, behaviour. For example, the intention to perform a particular task may be ascribed to a team (or collection) of agents, with no individual agent itself having the intention to perform that task. Ascribing such attitudes to teams of agents results in such concepts as *mutual beliefs, joint goals* and *joint intentions*. These concepts have been previously explored both formally and informally in the DAI literature [8, 11, 19, 20].

The attitudes of each agent may reflect the attitudes of the team, and vice versa. In our framework, the relationship between attitudes at different levels is specified via bridge rules. In this section, we explore the use of bridge rules for specifying team behaviours in terms of the behaviours of individual agents.

It should be noted that the ideas described in this section are at a rather preliminary stage. However, we believe there is scope for much interesting work in this area.

---

[9] Note that since we are dealing with multi-agent systems the state of Mary will depend on the timing of the events, i.e., whether the payment happens before or after Mary attempts to get money.

[10] For simplicity, we focus on these two modalities and ignore desire for the purposes of this paper.

6.1 Decomposing Attitudes

In our hierarchical, encapsulated model of contextual reasoning, concepts such as *mutual* belief and *joint* intention are not straightforward. Normally, a mutual belief is a belief which is held by every agent (in some collection of agents). However, we have allowed (indeed encouraged!) different agents to be concerned with different aspects of the world—if all agents deal with mutually disjoint languages, then there can be *no* belief which is held by all agents. However, if we consider the hierarchical model of agent specification, a belief at the system level may correspond to beliefs in each agent of that system—i.e., even if no agent can even express the belief $\mathsf{BEL}(\phi)$ (because ϕ is not part of its logical language), the system itself may be ascribed this belief which may in turn require that certain beliefs are held by the individual agents of that system.

Similar concerns relate to the other attitudes, particularly intention. For example, consider two agents *john* and *judy* who constitute a team. This team, call it *chef*, is given the intention to prepare dinner—i.e., the context for *chef* contains the theorem $\mathsf{INT}(prepare.dinner)$. Now *john* is a main-dish-preparing agent—i.e., *john* doesn't know how to do anything except prepare main dishes. On the other hand, *judy* is a dessert-preparing agent—*judy*'s function is to prepare desserts. In fact, *john* does not even know of the existence of desserts (i.e., *dessert* is not a term in *john*'s logical language), and similarly for *judy* and main-dishes. In order to achieve the *prepare.dinner* intention, separate intentions are lowered into the *john* and *judy* contexts—each of these lowered intentions are specified within the logical language understood by the corresponding agent. This could be done using the following bridge rules, which are associated with the *chef* system:

$$\frac{\mathsf{INT}(prepare.dinner)}{C_{john} : \mathsf{INT}(prepare.main.dish)} \qquad \frac{\mathsf{INT}(prepare.dinner)}{C_{judy} : \mathsf{INT}(prepare.dessert)}$$

Some debate has ensued as to whether joint attitudes can be completely decomposed into the attitudes of the component systems. For example, can a joint intention be seen as some sort of composition of the intentions of the individual agents, or is there some residual component of the joint intention that is not at all reflected in the agent intentions. The first approach can be referred to as a *reductionist* approach—the joint attitude can be completely reduced to the sum of the parts—and is supported by Grosz and Sidner [8] (amongst others). The second approach can be referred to as a *holistic* approach—the whole is greater than the sum of its parts—and is supported by Searle [20].

The framework of contextual specification that we have described above can be used to represent either view. For example, consider a two-level specification, i.e., where we have a collection of agent specifications (each with its own context C_i) and a context C corresponding to the team. Under the reductionist view, any intention in C would be reduced to intentions in C_i. Under the holistic view, there may be an intention in C which is not reduced to intentions in C_i. Furthermore, there may be beliefs and other formulas in C which are not reflected in any C_i.[11]

---

[11] Anand Rao (private communication) has pointed out that, in general, the holistic

6.2 Top-Down Specification of Intention

Given a BDI specification of agent behaviour, there are two main ways in which we can imagine attitudes, particularly intentions, being passed between agents and the system. In this section, we explore the alternative whereby intentions are passed from system to agent, i.e., *down* the hierarchy.

The specification of a system involves specifying the components (which may themselves be systems) and then defining the interactions as bridge rules. In a BDI architecture, agent behaviour is specified by the BDI theory, and particularly by how intentions arise and how the agent goes about achieving those intentions.[12] In the top-down specification model of a system, intentions are ascribed at the *team* level and the team then sets about achieving the intention, which may involve propagating various intentions down to the individual team members. For example, the *chef* system above is such a top-down specification. The *chef* system is given the intention of *prepare.dinner*—how the system comes to have this intention is not of immediate concern. The *chef* system itself may have no plans regarding how to achieve dinner—i.e., there is no plan library associated with the *chef* context. However, the bridge rules given earlier result in the *john* and *judy* systems each being assigned an intention—the first system is given an intention to prepare a main-dish, the second is given an intention to prepare a dessert. These systems may then have routines (i.e., plans) for achieving these particular intentions,[13] and will go about achieving their particular goals.

The question then arises as to how *chef*'s intention to prepare dinner is dropped. Assume that an agent drops an intention $\mathsf{INT}(\phi)$ when the agent has the belief $\mathsf{BEL}(\phi)$—i.e., an intention to achieve ϕ is dropped when ϕ is believed to hold. The agents *john* and *judy* would presumably drop their intentions when they respectively held the beliefs $\mathsf{BEL}(prepare.main.dish)$ and $\mathsf{BEL}(prepare.dessert)$. At the team level, it is clear that when these beliefs were held then the system could safely believe that dinner was prepared. This is captured by use of a lifting rule of the following form, associated with *chef*:[14]

---

stance can be effectively implemented by the reductionist stance by simply letting any residual attitude at the "joint" level effectively play the role of another agent, and consider the system of agents to include this virtual agent. For example, the context $\mathcal{C}$ would be incorporated as an "agent-level" context $\mathcal{C}_{n+1}$ and a new "team-level" context $\mathcal{C}'$ would be created. While this seems to be a possible approach, we simply want to illustrate that a team activity that is most naturally viewed holistically can be so modelled within our framework.

[12] This latter issue usually involves interaction with a plan library.

[13] Of course, *john* and *judy* may also be composed of agents/components to which they themselves pass down intentions.

[14] This rule requires a slight extension to the notation we have used so far, in that there are two different contexts involved in the premise. This could be avoided by having separate lifting rules which each raised a notification that the different courses had been prepared, and a rule internal to the *chef* context which inferred $\mathsf{BEL}(prepare.dinner)$ when both these internal conditions had been set.

$$C_{john} : \text{BEL}(prepare.main.dish) \wedge C_{judy} : \text{BEL}(prepare.dessert)$$
$$\overline{\text{BEL}(prepare.dinner)}$$

Once the belief in *prepare.dinner* had been raised to the *chef* context, the original intention would be dropped.

This view of top-down specification of behaviour promises many advantages. Each system behaviour could be independently specified using a BDI framework. At the team level, only the way in which complex intentions are to be delegated to the component system needs to be specified—i.e., there is no need for a plan library of how to achieve these more complex tasks. Of course, a BDI specification is still required at the team level—in particular, the way in which beliefs and intentions interact, and how team-intentions are dropped, needs to be defined. However, a natural view is that the beliefs at the team level are propagated up from the component agent level—i.e., the component agents perform tasks and monitor the world and (some) of these beliefs are lifted up to the team level.[15]

This top-down methodology for the specification of complex behaviour certainly seems to warrant further investigation.

6.3 Emergent Behaviour

Recent research in AI (and in the philosophy of mind) has taken the viewpoint that intelligent behaviour results from the complex interaction of simple behaviours.[16] This viewpoint is generally referred to as *emergent behaviour*—(seemingly) complex behaviour "emerges" from the interaction of simple behaviours. Our hierarchical framework seems to allow a natural method for modelling and reasoning about such behaviour.

Imagine a number of simple agents, each with its own limited view of the world and limited operations it can perform on its world. Assume the behaviour of each of these agents is specified by a BDI theory. Each agent would involve simple intentions and methods for achieving those intentions. In an "emergent" system, the interaction of these simple behaviours leads to a more complex behaviour at the team level. The difference between a system displaying emergent behaviour and one which is specified top-down is that the team-intention is not imposed on the system, but rather it *arises* from the interaction of the component systems. None of the team-members is aware of any team-view of intention—in fact, no team-member need be aware of the existence of a team, or even of any other agent. However, it is often convenient to be able to reason about the "intentions" of the team or collection of agents.

As a very simple example, consider again the agents *john* and *judy*. The *john* agent simply prepares main-dishes, with no knowledge of any wider task—the way in which *john* comes to obtain such intentions is not of our concern. Similarly, *judy* simply prepares desserts. We want to be able to reason about the

---

[15] Of course, when such beliefs are lifted, they need not be in the language of the agent. Further, it may only be that when some agent (or collection of agents) holds some belief, then some corresponding belief is lifted up to the team.

[16] For example, the work of Brooks [4] is notable in this respect.

behaviour of the overall system; in particular, we want to ascribe the collection of these two agents as having the intention to prepare dinners. This intention is ascribed to the team by use of the following lifting rule in $chef$:[17]

$$\frac{C_{john} : \mathsf{INT}(prepare.main.dish) \wedge C_{judy} : \mathsf{INT}(prepare.dessert)}{\mathsf{INT}(prepare.dinner)}$$

Once beliefs (and desires) of the agents have also been propagated upwards to the team level, an outsider can reason about the behaviour of the team (using a BDI theory), even though there is no real "team-oriented" behaviour as such— the individual agents simply go about their tasks and it is only as outsiders that we know (or care) about any team-oriented attitudes.

7   Concluding Remarks

In this paper, we have discussed a logical framework for the specification of multi-agent systems that incorporates several important design features from the object-oriented paradigm, namely: abstraction, encapsulation, modularity and hierarchy. This framework suggests a methodology for the formal specification of multi-agent systems whereby component systems may be independently specified since each agent need only be concerned with its own world. The final system is specified once the interactions between the component agents are defined. This leads to a methodology of true distributed design.

Recently, agent-specification via BDI architectures has gained popularity. This involves specifying the behaviour of an agent in terms of its beliefs, desires and intentions. Introducing a "team" aspect raises its own interesting issues relating to mutual beliefs, joint desires and joint intentions. We have explored the use of our distributed logical framework for the specification of team behaviour in terms of the behaviour of the component agents. In particular, a top-down specification leads to a view whereby the required team behaviour can be defined, which in turn results in agent behaviour that implements the team-intentions. Bottom-up specification, on the other hand, reflects a model of "emergent behaviour", whereby simple interacting agents give rise to a complex team-behaviour, which can then be reasoned about using the BDI theory at the system level.

There are a number of remaining technical issues that need to be resolved. Foremost of these involves the definition of a model-theoretic semantics for our syntactic framework. In particular, this should also allow us to address issues related to synchronization and temporal reasoning across systems of agents. Also, we wish to further explore the BDI specification of teams of agents, using the modal BDI theories that have appeared in the literature [6, 17, 18].

References

1. J. Barwise. *The Situation in Logic*. CSLI Lecture Notes 17, Stanford: CSLI Publications, 1989.

---

[17] There are issues to do with coordination that need to be resolved here.

2. G. Booch. *Object Oriented Analysis and Design with Applications*. The Benjamin/Cummings Publishing Company, Inc., Redwood City, California, 1993.

3. F. Brazier, B. D. Keplicz, N. R. Jennings, and J. Treur. Formal specification of multi-agent systems: a real-world case. In *Proceedings of the First International Conference on Multi-Agent Systems*, pages 25–32, San Francisco, 1995.

4. R. A. Brooks. Achieving Artificial Intelligence through building robots. Technical Report AI Memo 899, MIT, 1986.

5. A. Cimatti and L. Serafini. Multiagent reasoning with belief contexts II: Elaboration tolerance. Technical Report 9412-09, IRST, Trento, Italy, 1994.

6. P. R. Cohen and H. J. Levesque. Intention is choice with commitment. *Artificial Intelligence*, 42(3):213–261, 1990.

7. F. Giunchiglia, L. Serafini, E. Giunchiglia, and M. Frixione. Non-omniscient belief as context-based reasoning. In *IJCAI-93*, pages 548–554, Chambery, 1993.

8. B. J. Grosz and C. L. Sidner. Plans for discourse. In P. R. Cohen, J. Morgan, and M. E. Pollack, editors, *Intentions in Communication*, pages 417–444. MIT Press, 1990.

9. R. V. Guha. *Contexts: A Formalization and Some Applications*. PhD thesis, Department of Computer Science, Stanford University, Stanford, CA, USA, 1991.

10. P. Healey and C. Vogel. A Situation Theoretic model of dialogue. In K. Jokinen, editor, *Pragmatics in Dialogue Management*. Gothenburg Monographs in Linguistics, 1994.

11. D. Kinny, M. Ljungberg, A. Rao, E. Sonenberg, G. Tidhar, and E. Werner. Planned team activity. In *Proceedings of the Fourth European Workshop on Modeling Autonomous Agents in a Multi-Agent World*, Viterbo, Italy, 1992.

12. Y. Lesperance and H. J. Levesque. Indexical knowledge and robot action: a logical account. *Artificial Intelligence*, 72(1–2), 1995.

13. V. Lifschitz. Formal theories of action. In F. Brown, editor, *The Frame Problem in Artificial Intelligence*, pages 35–57. Los Altos, CA: Morgan Kaufman, 1987.

14. V. Lifschitz. Frames in the space of situations. *Artificial Intelligence*, 46:365–376, 1990.

15. J. McCarthy. Notes on formalizing context. In *IJCAI-93*, pages 555–560, Chambery, 1993.

16. D. Morley, M. Georgeff, and A. Rao. A monotonic formalism for events and systems of events. *Journal of Logic and Computation*, 4(5):701–720, 1994.

17. A. S. Rao and M. P. Georgeff. Modeling rational agents within a BDI-architecture. In J. Allen, R. Fikes, and E. Sandewall, editors, *Proceedings of the Second International Conference on Principles of Knowledge Representation and Reasoning*, pages 473–484. Morgan Kaufmann Publishers, San Mateo, CA, 1991.

18. A. S. Rao and M. P. Georgeff. BDI agents: from theory to practice. In *Proceedings of the First International Conference on Multi-Agent Systems*, pages 312–319, San Francisco, 1995.

19. A. S. Rao, M. P. Georgeff, and E. A. Sonenberg. Social plans: a preliminary report. In Y. Demazeau and E. Werner, editors, *Decentralized Artificial Intelligence: Volume 3*, pages 57–76. Elsevier, Amsterdam, 1992.

20. J. R. Searle. Collective intentions and actions. In P. R. Cohen, J. Morgan, and M. E. Pollack, editors, *Intentions in Communication*, pages 401–415. MIT Press, 1990.

A Kernel-Oriented Model for Autonomous-Agent Coalition-Formation in General Environments *

Onn Shehory and Sarit Kraus

Department of Mathematics and Computer Science
Bar Ilan University Ramat Gan, 52900 Israel

Abstract. An important way for autonomous agents to execute tasks and to maximize payoff is to share resources and cooperate on task execution by creating coalitions of agents. Among individual rational agents, such coalitions will form if, and only if, each member of a coalition gains more if it joins the coalition than it could gain previously. There are several models of creating such coalitions and dividing the joint payoff among the members. In this paper we present a model for coalition formation and payoff distribution in general environments. We focus on a reduced complexity kernel-oriented coalition formation model. The model is partitioned into two levels – a social level and a strategic level. This partition enables one to distinguish between regulations that must be agreed upon and are forced by the designers of the agents, and strategies by which each agent acts and that can be adopted at will.

1 Introduction

An important method for cooperation in multi-agent environments is coalition formation. Membership in a coalition may increase the agent's ability to satisfy its goals and maximize its own personal payoff. Game theory literature such as [11] describes which coalitions will form in N-person games under different settings and how the players will distribute the benefits of the cooperation among themselves. These results do not take into consideration the constraints of a multi-agent environment, such as communication costs and limited computation time, and do not present algorithms for coalition formation. Our research presents a multi-agent approach to the coalition formation problem, and provides a coalition-formation procedure. The paper deals with autonomous agents, each of which has tasks it must fulfill and access to resources that it can use to fulfill these tasks. Agents can satisfy goals by themselves, but may also join together to satisfy their goals. In such a case we say that the agents form a coalition.

In this paper we present a modification of the Kernel concept from game theory [3]. The modified Kernel serves as a basis for a polynomial-complexity mechanism for coalition formation. The mechanism is partitioned into two levels

---

* This material is based upon work supported in part by the NSF, grant No. IRI-9423967 and the Israeli Ministry of Science, grant No. 6288. We thank S. Aloni and M. Goren for their major contribution to the implementation of the model.

- the social level and the strategic level. The coordination-regulation protocols constitute the social level[2]. Different designers of agents must agree upon the regulation protocols of the social level in advance. The strategic level consists of strategies for the individual agent to act in the environment for maximization of its own expected payoff, given the social level, and can be decided upon by individual agents during the coalition formation process.

2 Environment Description

In order to design a coalition formation algorithm that will lead to the creation of mutually-beneficial coalitions, we give some general notations and definitions for concepts. We make the following assumptions:

- Various communication methods exist, so that the agents can negotiate and make agreements [14]. However, communications require time and effort.
- Resources can be transferred among agents in the environment. We also assume that there is a monetary system (e.g, money or utility points [7]) that can be used for side-payments. The agents use this monetary system in order to evaluate resources and productions that result from the use of the resources. The money is transferable among the agents and can be redistributed in a case of coalition formation.

The monetary system is part of the regulation of the environment. We present it as a possible regulation since it increases the agents' benefits from cooperation, although agents may reach agreements and form coalitions even if the second assumption is not valid (e.g., [6, 1]).

Below, we adjust some definitions from game theory to DAI environments. We consider a group N of n autonomous agents, $N = \{A_1, A_2, \ldots, A_n\}$. The agents are provided with, or have access to, resources. Each agent A_i has its own resource vector $q_i = \langle q_i^1, q_i^2, \ldots, q_i^l \rangle$, which denotes the quantities of resources that it has. The agents use the resources they have to execute their tasks. A_i's outcome from task-execution is expressed by a payoff function from the resources domain Q to the reals. Such a function $U^i : Q \to \mathcal{R}$ exchanges resources into monetary units. Each agent tries to maximize its personal payoff.

Individual self-motivated agents can cooperate by forming coalitions. A coalition is defined as a group of agents that have decided to cooperate and have also decided how the total benefit should be distributed among them. Formally:

Definition 1. Coalition Given a group of agents N and a resource domain Q, we define a coalition as a quadrate $C = \langle N_C, q_C, \overline{Q}, U_C \rangle$, where $N_C \subseteq N$; $q_C = \langle q^1, q^2, \ldots, q^l \rangle$ is the coalition's resource vector, where $q^j = \sum_{A_i \in N_C} q_i^j$ is the quantity of resource j that the coalition has. $\overline{Q}$ is the set of resource vectors after the redistribution of q_C among the members of N_C ($\overline{Q}$ satisfies $q^j = \sum_{A_i \in N_C} \overline{Q}_i^j$). $U_C = \langle u^1, u^2, \ldots, u^{|C|} \rangle$, $u^i \in \mathcal{R}$, is the coalitional payoff vector, where u^i is the payoff of agent A_i after the redistribution of the payoffs.

---

[2] We use the concepts of the social level and the notion of regulations interchangeably.

Definition 2. Coalition Value Let $C = \langle N_C, q_C, \overline{Q}, U_C \rangle$. V is the value of C if the members of N_C can together reach a joint payoff V. That is, $V = \sum_{A_i \in N_C} U^i(\overline{Q}_i)$, where U^i is the payoff function of agent A_i and $\overline{Q}_i$ is its resource-vector after redistribution in C.

This coalition value provides the agents with a method for evaluating coalitions. Note that the specific distribution of the resources among the members of the coalition strongly affects the results of the payoff functions of the agents and thus affects the coalitional value. It is in all of the agents' interest to reach a resource distribution that will maximize the coalitional value. Therefore, the resources are redistributed within $\overline{Q}$ in a way that maximizes the value of the coalition. Thus, the coalition value V of a specific group of agents N_C is unique. The complexity of computing the redistribution of the resources and calculating the coalitional value depends on the type of payoff function of the coalition-members. For example, if the payoff functions are linear functions of resources, then the simplex method can be applied (this specific method is polynomial). If the payoff functions are not linear, it is not guaranteed that the agents will be able to calculate the value of a coalition within a polynomial time. Another important assumption that we make about the coalition value is that the value of a given coalition does not depend on the other coalitions that are formed.

Assumption 1 Coalition joining (personal rationality)
An agent joins a coalition only if it can benefit at least as much within the coalition as it could benefit by itself. An agent benefits if it fulfills some or all of its tasks, or gets a payoff that compensates it for the loss of resources or non-fulfillment of some of its tasks[3].

Coalition-formation usually requires disbursement of payoffs among the agents We define a payoff vector $U = \langle u^1, u^2, \ldots, u^n \rangle$ in which its elements are the payoffs to the agents. In each stage of the coalition formation process, the agents are in a coalitional configuration. That is, the agents are arranged in a set of coalitions $\mathcal{C} = \{C_i\}$, that satisfies the conditions $\cup_i C_i = N$, $\forall C_i, C_j$, $C_i \neq C_j$, $C_i \cap C_j = \phi$. A pair of a payoff vector and a coalitional configuration are denoted by $PC(U, \mathcal{C})$, or just PC (Payment Configuration). Since we assume individual rationality of the agents, we consider only individually rational payment configurations (which are denoted as IRPC's in the game theory, e.g.,[9]). We define a coalitional configuration space (CCS) and a payment configuration space (PCS). A (rational) CCS is the set of all possible coalitional configurations such that the value of each coalition within a configuration is greater or equal to the sum of the payoffs of the coalition-members[4]. The size of the CCS can be proven to be $O(n^n)$ using standard combinatorial methods. A PCS is a set of possible individually-rational PCs. That is, a PCS consists of pairs (U, C) where U is an individually-rational payoff vector and C is a coalitional

---

[3] This assumption is usually called "personal rationality" in the game theory [11, 8].
[4] It is very common to normalize the agents' payoffs to zero. In such a case, the requirement on the sum becomes simpler – the coalitional values must be non-negative.

configuration in CCS. Since for each coalitional configuration there can usually be an infinite number of payoff vectors, the number of PCs is infinite, and the PCS of all rational PC's is an infinite space.

We would like the resulting payoff vector of our coalition-formation model to be stable and Pareto-optimal. A payoff vector is Pareto-optimal if there is no other payoff vector that dominates it, i.e., there is no other payoff vector that is better for some of the agents and not worse for the others [8]. It seems in the best interest of individually rational agents to seek Pareto-optimal payoff vectors. However, a specific Pareto-optimal payoff vector is not necessarily the best for all of the agents. There can be a group of Pareto-optimal payoff vectors where different agents prefer different payoff vectors. This may lead to difficulty when agents negotiate cooperation and coalition formation. Therefore, Pareto-optimality is not sufficient for the evaluation of a possible coalition for a specific agent, and we shall present the concept of stability.

The issue of stability was studied in the game theory literature in the context of n-person games [11, 8]. The notions of stability they developed are useful for our purposes, when coalitions are formed during the coalition formation procedure. The members of such coalitions can apply these techniques to the distribution of the coalitional value. Game theorists have given several solutions for n-person games, with several related stability notions. In this paper we concentrate solely on the Kernel solution concept. However, we shall discuss the other solution concepts in an extended version of this paper.

3 The Kernel K

The kernel [3] is a PCS in which the coalitional configurations are stable in the sense that there is equilibrium between pairs of individual agents which are in the same coalition. Two agents A, B in a coalition C, in a given PC, are in equilibrium if they cannot outweigh one another from C, their common coalition. Agent A can outweigh B if A is stronger than B, where strength refers to the potential of agent A to successfully claim a part of the payoff of agent B in PC.

During the coalition formation, agents can use the kernel solution concept to object to the payoff distribution that is attached to their coalitional configuration. This objection will be done by agents threatening to outweigh one another from their common coalition. The objections that agents can make given a $PC(U, C)$ are based on the excess concept. We recall the relevant definitions.

Definition 3. Excess The excess [3] of a coalition C with respect to the coalitional configuration PC is defined by $e(C) = V(C) - \sum_{A_i \in C} u^i$, where u^i is the payoff of agent A_i in PC. C is not necessarily a coalition in PC, and it can be in any other coalitional configuration. $V(C)$ is the coalitional value of coalition C as in definition 2.

The number of excesses is an important property of the kernel solution concept. Given a specific PC, the number of the excesses with respect to the specific coalitional configuration is 2^n. Any change in the payoff vector U, either when the coalitional configuration changes or when it remains unchanged, may cause a change in the set of excesses. Such a change will require recalculation of all of the excesses. Agents use the excesses as a measure of their relative strengths. Since a higher excess correlates with more strength, rational agents must search for the highest excess they have. This maximum is defined by the surplus.

Definition 4. Surplus and Outweigh The maximum surplus S_{AB} of agent A over agent B with respect to a PC is defined by $S_{AB} = MAX_{C|A \in C, B \notin C} e(C)$, where $e(C)$ are the excesses of all the coalitions that include A and exclude B, and the coalitions C are not in PC, the current coalitional configuration. Agent A outweighs agent B if $S_{AB} > S_{BA}$ and $u^B > V(B)$, where $V(B)$ is the coalitional value of agent B in a single agent coalition.

In other words, given a coalitional configuration and a payoff vector, the agents compare their maximum surpluses, and the one with the larger maximum surplus is stronger[5]. The stronger agent can claim a part of the weaker agent's payoff in the same coalitional configuration, but this claim is limited by the personal rationality. The personal rationality is expressed by the requirement $u^B > V(B)$, which means that in any suggested coalition, agent B must receive more payoff than it gets by itself in a single-member coalition. Therefore, agent A cannot claim an amount that would leave agent B with $V(B)$ or less. If two agents cannot outweigh one another, we say that they are in equilibrium. We say that A, B are in equilibrium if one of the following conditions is satisfied:
1. $S_{AB} = S_{BA}$; 2. $S_{AB} > S_{BA}$ and $u^B = V(B)$; 3. $S_{AB} < S_{BA}$ and $u^A = V(A)$.
Using the concept of equilibrium, the kernel and its stability are:

Definition 5. Kernel and K-Stability
A PC is K-stable if $\forall A, B$ agents in the same coalition $C \in PC$, the agents A, B are in equilibrium. A PC is in the kernel iff it is K-stable.

The kernel stability concept provides a stable payoff distribution for *any* coalitional configuration in the CCS[6]. Using this distribution, the agents can compare different coalitional configurations. However, checking the stability does not direct the agents to a specific coalitional configuration. The coalition formation model that we develop will perform this direction. The kernel is a solution that always exists for all of the coalitional configurations in the rational CCS [3, 2]. In addition, the kernel leads symmetric agents to receive equal payoffs. That is, agents with the same bargaining strength (which is expressed by identical sets of excesses) will gain equal payoffs. Such symmetry is not always guaranteed in other solution concepts (e.g., the bargaining set) [11]. Another property of the kernel – it is a comparatively small subset of the PCS of all rational PC's. In

---

[5] Note that the surpluses and the excesses refer to payoffs and are expressed in monetary units. Therefore they can be compared.

[6] The kernel does not provide coalitional configurations, it only determines how the payoffs will be distributed given a coalitional configuration. Therefore, it cannot be used as a coalition formation algorithm.

addition, the mathematical formalism of the kernel allows one to divide its calculation into small processes, thus simplifying it. Some exponentially-complex computing schemes for the kernel solution were provided, e.g., by [2]. Stearns [13] presented a transfer scheme that, given a coalitional configuration and a payoff vector, converges to an element of the kernel. Due to its advantages, the kernel was chosen as the theoretical background for our coalition formation model.

4 Coalition Negotiation-Oriented Algorithm (CNA)

We shall present here the CNA, a coalition formation algorithm which is based on negotiation [5] among the agents. The CNA consists of steps in which coalitions transmit, accept and reject proposals for creating new coalitions. The CNA will start with all agents being in single-membered coalitions, and will proceed through a sequence of CNA steps as described in 4.2. In each step, at least one coalition will make an attempt to improve the payoffs of its members by making a coalition formation proposal to another coalition. The acceptance of such a proposal will improve the situation of the agents involved. The CNA may continue either until all of the proposals of all of the agents are rejected or until a K-stable and Pareto-optimal PC has been reached, thus reaching a steady state. The CNA may also terminate when the time-period that was allocated ends.

When the agents use the CNA, the coalitional configurations that are formed when the agents reach the steady state are stable according to a new stability concept that we define below, the polynomial-K-stability. However, since the CNA is an anytime algorithm [4], even if it is terminated after a limited number of steps *before* reaching a steady state, it will still provide the agents with a polynomial-K-stable PC.

4.1 The Polynomial Approach

We shall present some modifications of concepts to adjust them to the polynomial-K-stability algorithm. We define polynomial excesses as excesses that are calculated using a polynomial set of coalitions. That is, excesses which are calculated with respect to a polynomial subset of all 2^n possible coalitions. The designers of agents must agree upon regulations that will direct their agents to a well defined polynomial set of coalitions. In order to reach such a subset, we suggest that the designers agree upon two integral constants K_1, K_2. These constants $K_1 \le K_2$ should not depend on n. Nevertheless, K_1, K_2 that are small compared to $n/2$ should be preferred[7] (see section 4.4). In the regulation of the coalition formation model, the agents shall be allowed to consider excess calculations only for coalitions of sizes in the ranges $[K_1, K_2]$. Choosing K's contradictory to the agreed upon K_1, K_2 by a specific agent cannot guarantee, in advance, an improvement in its payoff, and therefore should be avoided. In addition, according to the regulations, objections based on different K's are not acceptable. The

---

[7] There are occasions where the properties of the system confine the coalition sizes.

constants K_1, K_2 can be determined via simulation. We have developed a simulation tool and shall discuss it later in this paper. Since the agents will compute only polynomial excesses in order to outweigh one another, we shall introduce the polynomial surpluses and the polynomial K-stability:

Definition 6. A polynomial maximum surplus SP is a maximum surplus that is computed from a set of polynomial excesses. A coalition C in a coalitional configuration is polynomially-K-stable if for each pair of agents $A, B \in C$, either one of the agents has a null payoff in C, or $SP_{AB} - SP_{BA} \leq \varepsilon$, i.e., the agents are in equilibrium with respect to ε, where SP are the polynomial surpluses, and ε is a small predefined constant.

Given a specific coalitional configuration with an arbitrary payoff distribution vector, it is possible to compute a polynomial-K-stable PC by using a truncated modification of the convergent transfer scheme in [13]. We implement the Stearns-scheme by using the n-correction of Wu [15][8] to initialize the process. The iterative part of the scheme is modified so that it will terminate whenever a payoff vector that is close, according to the predefined small ε, to an element of the polynomial kernel has been reached.

4.2   A CNA Scheme

The CNA scheme is aimed at advancing the agents from one coalitional configuration to the other, to achieve more cooperation and increase the agents' payoffs. Given a specific coalitional configuration, the agents try to find a correspondent *stable* payoff vector. The social level of the CNA will be constructed as follows:

Regulation 4.1 Negotiation scheme

1. *At the beginning of the coalition formation, all entities are single agents.*
2. *In the first stage of the CNA, members of a coalition may receive proposals only as part of the coalition (thus, coalitions can only expand in this stage).*
3. *Each coalition in the first stage of the CNA will coordinate its actions either via a representative or by voting (or both) e.g., [10]. A representative will be responsible for coalitional actions. We omit the strategic details of the coalitional representative and voting due to space limitations.*
4. *In this stage, the coalitions iteratively perform the following sequence of steps:*
   - *Each coalition, C, will decide, according to its specific strategy, with which other coalitions it is interested in forming a joint coalition.*
   - *Each coalition will design proposals to be offered to others (details in 4.3) and transmit one proposal to one target coalition at a time. After having transmitted a proposal, the coalition must wait for response (see 4.3).*

---

[8] While Wu uses this correction as a means for iterating in a transfer scheme for the core, we use it as a single correction in the beginning of our iterative algorithm. This n-correction is not necessary in the Stearns scheme.

     — *In cases where the offer was accepted, the offering coalition and the accepting coalition will join together, thus forming a new coalition according to the details of the proposal. In cases where the agents act according to the representative approach, the members of the newly-formed coalition shall decide who the coalitional representative will be[9].*

     — *The new PC, with the new payoff vector that results from the acceptance of a proposal by a coalition, determines the payoffs of the agents. Agents that are dissatisfied with their new payoff can only react in the next iteration via their coalition.*

     — *In case other proposals were active (that is, transmitted but not yet answered), these proposals will be canceled. After the acceptance of the specific new PC and the disbursement of the payoffs (to all of the agents in the environment) according to the corresponding payoff vector, no further changes in the coalitions take place in the current iteration.*

5. *This sequence of steps shall be repeated until a steady state is reached.*

6. *To enable coalitions to perceive that a steady state has been reached, each coalition shall announce its status (i.e., that it has no more proposals to transmit). If all coalitions announce that they have no more proposals to make, then they reach a steady state. The announcement of a coalition regarding its status is valid only for the current iteration. If the agents run out of computation time before a steady state has been reached, the algorithm terminates and the last PC holds.*

7. *In the second stage of the CNA, the coalitions will follow the same sequence of steps as in the first stage of the CNA. However, proposals that involve destruction are allowed.*

8. *In this stage, when a new PC, with the new payoff vector that results from the acceptance of a proposal by a coalition, changes the payoffs of the agents, it may lead agents to leave their coalitions. These coalitions will destruct.*

9. *The second stage will end either when a steady state[10] is reached (as in part 3 of this regulation) or when the computation time ends.*

10. *To enable the perception that a steady state has been reached, each coalition shall announce its status when it has no more proposals to transmit, independently from such announcements in the first stage.*

The limitation that destruction of coalitions be avoided in the first stage will radically shorten the coalition formation process by avoiding most of the intra-coalitional computation and communication in this stage. The number of iterations for reaching a steady state in the first stage of the CNA is $O(n)$. If the agents have enough time and computational resources and did not reach a PC (which is polynomially-K-stable and locally Pareto-optimal[11]) during the first stage, and the CNA was not halted by time constraints, the agents will

---

[9] The payoff vector U of the new PC is valid from now on and is used as the basis for future negotiation.

[10] Note that the steady state in the second stage is different from the steady state in the first stage since there are different restrictions on the proposals in these two stages.

[11] We denote Pareto-optimality as local when it is examined w.r.t a subset of the CCS.

continue through stage 2. The number of steps until such a PC is reached is greater than $O(n)$ and may be $O(n^n)$ due to the size of the PCS. As a strategy for recognizing local Pareto-optimal PC's by the agents[12], we recommend that an agent will compare all of the PC's that were computed through the CNA iterations. A comparison of two PC's entails comparing the corresponding payoffs of all agents.

It was stated above that an acceptance of a proposal implies an acceptance of the corresponding payoff vector by all of the agents. This acceptance may change the payoffs of some of the agents even when they are not involved in the negotiation. This can happen because a proposal consists of a payoff vector with n elements which are the payoffs of the n agents. The aim of the change in the payoffs is to preserve the polynomial-K-stability. Since some of the agents may be dissatisfied with the new payoff they receive, they can make new proposals according to which they receive a greater payoff. If their proposal is accepted they will improve their situation.

We suggest that a coalition C_p that has to decide which other coalitions to approach, as required according to regulation 4.1, use the following strategies:

Strategy 4.1 Choosing and directing proposals

- *Coalition C_p will rank the other coalitions according to the payoffs that C_p expects that its members will receive from forming a joint coalition with one of the other coalitions. This ranking can be done in the following ways:*
    - *by explicitly calculating the value of the joint coalitions and the K-stable corresponding payoff vector (such calculations are almost as complex as designing complete proposals for all other coalitions).*
    - *by making an estimate of the expected payoffs. This can be done by checking the quantities of resources that the other coalitions have and examining the previous payoffs that the coalition-members have received.*

    C_p will choose one of these options according to its computational resources.
- *The coalition C_p will compute (possibly via its representative) under the polynomial restrictions, the proposal that is expected to bring it maximum payoff if coalition C_r, the receiving coalition, will accept it. The details of this calculation are in strategy 4.2.*
- *Next, a possible step in the second stage of the process but not an obligatory one, is that coalition C_p will compute proposals that do not give maximum payoff immediately, but lead, through more than one CNA step, to future coalitions that will provide the members of C_p with more payoff than the first proposal brings them.*

The more-than-one-step-forward approach may lead to improved results, but meanwhile it radically increases the complexity of the algorithm.

---

[12] The recognition of local Preto-optimality is necessary in the second stage of the CNA.

4.3 CNA Regulation-Details and Strategies

Proposal Structure and Design According to part 4 of regulation 4.1, proposals for the generation of new coalitions should be designed by the current coalitions. We denote a coalition that designs and transmits a proposal by C_p and a coalition that receives a proposal by C_r. Other coalitions will be denoted by C_a. A coalition formation proposal must have the following properties:

Regulation 4.2 Proposal *A proposal is a full description of the proposed coalition and coalitional configuration. It shall be polynomially-K-stable as described in section 4.1. To satisfy all agents' personal rationality, coalition C_p will design only new coalitions C_{new}, $N_{new} = N_p \cup N_r$, with payoff vectors U_{new} that give the members of C_{new} higher payoffs than they receive in their current coalitions C_p and C_r. In addition to the newly-designed coalition C_{new}, C_p will design the coalitional configuration in which C_{new} is included. The membership of agents in the coalitions that are not involved in the current coalition formation proposal stay unchanged in the newly-designed coalitional configuration. This coalitional configuration with the modified payoffs is the proposal that C_p shall send to C_r.*

We suggest that C_p, that designs a proposal for C_r, use the following strategy:

Strategy 4.2 Proposal design *Coalition C_p shall calculate the coalitional value of the joint coalition V_{p+r}. If the condition $V_p + V_r < V_{p+r}$ is not satisfied, then a proposal will not be generated. In such a case C_p shall stop the process of designing a proposal for C_r. Otherwise, $(V_p + V_r < V_{p+r})$, C_p shall calculate the coalitional values of all other coalitions of all sizes in the range $[K_1, K_2]$. C_p shall calculate the payoff vector U_{new} of the new PC wherein coalitions C_p and C_r join to form C_{new}, and all the other coalitions do not vary. These calculations will be done by using the truncated transfer scheme, starting from the initial payoff vector U. Coalition C_p will compare U_{new} to U. If the payoffs to all of the members of C_r and C_p in PC_{new} are greater or at least equal to their payoffs in the current PC and is also better than all of the proposals that C_p has in the received-proposals queue (see strategy 4.3) then coalition C_p will send the resultant PC_{new} as a proposal to coalition C_r. Otherwise, C_p shall stop the process of designing a proposal for C_r.*

This strategy for proposal design shall be used by agents that are interested in reaching beneficial coalition formation and act under the regulation 4.2, which forces polynomial-K-stability of proposals. This is because the calculation of the new coalitional value V_{p+r} and the comparison to the sum of original coalitional values V_p and V_r is done to avoid worthless proposals in advance. In cases where V_{p+r} enables beneficial coalition joining, coalition C_p shall seek all coalitional values (of coalitions of sizes in the range $[K_1, K_2]$) in order to use these values for PC calculations.

Receiving Proposals and Responding Coalitions may either design and transmit proposals or receive proposals (or both). We suggest that a coalition C_r that received a proposal from C_p will act as follows:

Strategy 4.3 Proposal evaluation and response

- C_r shall hold a queue of the proposals that it has received and not yet answered. The proposals will be held in the queue until their predefined time-period[13] has expired.
- When a new proposal from C_p is received by C_r, the payoffs to the members of C_r according to the new proposal should be compared to their current payoffs. If some of the members of C_r receive in C_{new} less payoff than they receive in C_r, the proposal should be rejected immediately. Otherwise, C_r shall insert the proposal into the queue.
- In order to decide which proposal to accept from the queue, C_r shall find the one that gives its members the largest payoff increments via payoff vector comparison. Note that different agents may prefer different proposals. Hence, the assessment and agreement upon a proposal may be resolved according to the coalition decision-making method. In a case where more than one proposal is preferred equally, C_r shall act according to the time-ordering regulation (see section 4.3).
- If C_r is not obliged to honor a proposal and the time period of the best proposal in the queue is just about to expire, then C_r shall accept this proposal.
- C_r sends an acceptance message to the coalition C_p, that its proposal was accepted, and rejection messages to all others who have made proposals.

Above, we suggest that C_r will evaluate the received proposal with respect to its current state and to other proposals. We do so even though we expect the proposals to be better than the current state of C_r. This expectation is due to the assumption that the agents follow our strategy for designing proposals. However, agents are allowed to use other strategies[14].

In the strategy that we have presented above, we proposed that a coalition that received a proposal should accept it only when the time period is about to expire. This is because waiting until the last moment to respond increases the probability for the acceptance of more proposals that may be better than those that are already in the queue.

Proposal Ordering and Preventing Deadlocks The social level of the CNA shall include a method for determining the priorities of proposals. Agents that are members of coalitions that receive these proposals should use this method on occasions when they have received more than one proposal wherein their payoffs are the same. This is because proposal priority may influence the results of the CNA, and different ordering methods will cause different coalitions and different payoff distribution.

It may happen that while coalition C_p is waiting for C_r to respond, it receives a better proposal from a coalition C_a. It can happen that C_r sent a proposal to C_a and delayed its answer to C_p because of this proposal. Such a situation can

---

[13] To avoid unlimited waiting for response, a constraining time-period shall be defined.
[14] Strategies are only suggestions, while regulations are laws that are enforced.

cause deadlocks. In order to prevent this, a coalition C_p shall act according to a deadlock-preventing regulation. The regulations for proposal ordering and for preventing deadlocks are not presented in this paper due to space restrictions.

4.4 Complexity of the CNA

The complexity of the CNA is strongly affected by the difference between K_1 and K_2. According to these constants, the agents compute the polynomial set of coalitions, the coalitional values and the coalitional configurations. The complexity of these calculations is of the same order of the number of the coalitions. The number of coalitions in the CNA is given by

$$n_{coalitions} = \sum_{i=K_1}^{K_2} \binom{n}{i} = \sum_{i=K_1}^{K_2} \frac{n!}{i!(n-i)!}$$

which is a sum of polynoms of order $O(n^i)$. Summation over $K_2 - K_1$ polynoms is a polynom of the largest order of all polynoms in the sum.

Computation of coalitional values and configurations

The CNA requires the computation of $n_{coalitions}$ coalitional values. It also requires the design of coalitional configurations, and the number of these depends on the time constraints. In a case where only the first stage of the CNA is performed, the number of coalitional configurations which are treated is $O(n)$. If the CNA proceeds through the second stage, the number of coalitional configurations increases. In each iteration of the CNA, when one coalitional configuration is treated, a polynomial-K-stable PC shall be calculated.

Computation of polynomial-K-stable PC's

The CNA will employ the transfer scheme for calculating polynomial-K-stable PC's. Each iteration of the transfer scheme is constructed from the following:

- For any given payoff vector, there are $n_{coalitions}$ excesses. All of these excesses are calculated in each step, and the calculation of each excess is of the order of n (see definition 3). These excesses are searched for surpluses. The excess and surplus calculations are together of order $O(n \times n_{coalitions})$.
- Among the surpluses, the greatest is found by $O(n_{coalitions})$ operations.
- Some additional calculations of a lower complexity are required to complete one iteration of the transfer scheme.

The total complexity of one iteration of the transfer scheme is $O(n \times n_{coalitions})$. The number of iterations that should be performed to reach convergence depends on the predefined allowed error ε. The resulting payoff vector of the transfer scheme will converge to an element of the polynomial-kernel (with a relative error not greater than ε) within $n \log_2(e_{r_0}/\varepsilon)$ iterations [13], where e_{r_0} is the initial PC relative error.

The transfer scheme will be performed for $O(n)$ coalitional configurations in the first stage of the CNA and up to $O(n^n)$ in the second stage. Therefore, the complexity of the CNA is of at least $O(n^2 n_{coalitions})$ and up to $O(n^n n_{coalitions})$ computations. If the computations are distributed among the agents, this order of complexity of computations is divided by n. There is an additional communicational complexity, which is of order $O(n^2 n_{coalitions})$.

5   Example and Implementation

As an example, we present a specific automated taxi-drivers domain consisting of four taxi-drivers (Ann, Barbara, Christie and Debbie). Each owns a black-taxi and located in Victoria station, London and uses an automated agent for planning. Each black-taxi can carry up to 4 passengers. The taxi-drivers' costs are 30 pence per mile, an insurance fee of 1 pound per day per taxi or 1.50 pounds per day per two taxis. The drivers' income is composed of a 60-pence base rate per trip and 80 pence per mile. The profit p from one trip is $p = 0.6 + 0.5m$, (m is the number of miles). After every trip, the drivers return to Victoria station. It may be in each driver's interest to cooperate and form coalitions in order to increase her transportation capabilitiesand reduce the insurance costs.

The agents have received the following tasks: Ann has to take 7 people from Victoria station to Sussex gardens; Barbara has to take 1 passenger from Victoria station to Edgeware road; Christie has to take 6 passengers from Kensington gate to Victoria station; Debbie has to take 1 passenger from Oxford street to Victoria station. The distances between places are: $d(V,S) = 4$; $d(V,O) = 3$; $d(V,E) = 5$; $d(V,K) = 4$; $d(S,O) = 1$; $d(S,E) = 1$; $d(S,K) = 1$; $d(K,O) = 2$; $d(K,E) = 2$; $d(O,E) = 2$. A taxi-driver coalition payoff function is of the form $U = \sum(0.6 + 0.5m_{trans}) \times \lceil psg/4 \rceil + \sum 0.3 m_{parl} - Ins(cabs)$, where the sums are over the tasks of the members of the coalition, m_{trans} is the number of miles per trip (back and forth), m_{parl} is the number of miles that can be avoided via cooperation and, $\lceil psg/4 \rceil$, is the number of trips that a single taxi must perform due to its number of passengers psg. $Ins(cabs) = 0.75 \times cabs + 0.25 \times (cabs \bmod 2)$, the insurance costs, is a function of $cabs$, the number of taxis in the coalition. For example, the cooperation of Ann and Christie in a coalition AC will enable them to avoid 14 miles of driving. Therefore $m_{parl} = 14$, and $V(AC) = (2(0.6 + 0.5 \times 8) + (0.6 + 0.5 \times 6)) + (0.3 \times 14) - 1.5 = 21.1$. Other coalitional values are: $V(A) = 8.2; V(B) = 4.6; V(C) = 8.2; V(D) = 2.6; V(BD) = 9.5, V(AB) = 13.3$. An explicit exponential calculation of k-stable Pareto-optimal PC's may result in several coalition configurations, e.g., (AC, BD) or (ABCD). For the first configuration, a stable Payoff vector will be, e.g., $U = (11.4, 4.6, 9.7, 4.6)$.

We have developed a simulation tool to examine the performance of the CNA, and have run the simulation tool for the example above. The first coalition that was formed[15] was AC and the payoff vector was $U_1 = (11.4, 4.6, 9.7, 2.6)$, which is polynomially k-stable but not k-stable and not Pareto-optimal. Therefore, the negotiation continued and the next coalition was BD, and the payoff vector was $U_2 = (11.4, 4.6, 9.7, 4.6)$. U_2 is k-stable and Pareto-optimal, and agrees with the exponential-calculation results. The Pareto-optimal simulation-result was achieved both with limited K_1 and K_2 and without this limitation. The

---

[15] The suggestions' order bases on the ranking heuristics, according to which Ann prefers Christie over the others, and vice versa. Hence, AC is a preferred suggestion.

CNA terminated normally with this (AC, BD) result, although (ABCD) is also Pareto-optimal, because (ABCD) does not improve the payoffs to all agents, and the CNA was designed to avoid coalition formation that does not improve the payments. These results were obtained for all runs. However, the CNA is not deterministic, and its results may depend on the order of offering during the negotiation. For example, when we increased the allowed error ε (this increment reduces the computational complexity) in some of the runs, another PC was reached: (A, BC, D), which is polynomially k-stable, but not Pareto-optimal.

The performance of the CNA was tested with respect to different constants (K's,ε) and different environmental settings. For algorithmic and computational-complexity reasons, the implementation can process only polynomial payoff functions with the assumption of independent resources. In most of the examples we concentrate on linear payoff functions. In our experiments, the computational complexity led us to choose the constants K_1 and K_2 to be small, usually in the range of 1 to 5. We have also allowed the simulation to run with no K limitation, to reach the exponential solution when the number of agents is small (l.t. 10).

Running the simulation has provided only preliminary results. We shall report more results in future work. To date, we have shown that the simulated CNA reaches a stable PC within a reasonable time (for small K's, and only for the first stage of the CNA). In addition, it has been found that for the settings that we have examined, the CNA continuously improves the agents' payoffs, whether it is normally terminated or halted artificially. According to these preliminary results, we tend to believe that the CNA shall be proven to be a good coalition formation model for MA systems in general environments.

6 Conclusion

The coalition formation model (CNA) we present is negotiation-oriented. It is usfull for instances where the number of agents may be large (e.g., tens of agents), computations are costly and time is limited. This is because the model leads to coalition formation within a polynomial time and a polynomial amount of calculations. The ability to reach such complexity is due to a new approach that we present for developing this model. We introduce the new concept of *polynomial* K-stability. The original K-stability refers to a PC where agents cannot make justified objections, using excess and surplus calculations, against members of the same coalition with respect to their payoffs in the PC. In general, the computation of either objections or counter-objections is of exponential complexity. In cases where the agents have neither the time nor the computational ability to compute these objections and counter-objections, stability can be reached by concentrating on polynomial objections. The new polynomial K-stability entails the calculation of objections in polynomial complexity.

The CNA leads to distribution of both calculations and communications. Another advantage of the CNA is that it is an anytime algorithm: if halted after any negotiation step, it provides the agents with a set of formed polynomial-K-stable coalitions. A deficiency of the CNA is that in polynomial time it cannot guarantee that a Pareto-optimal PC will be reached. However, for calculating a Pareto-optimal PC all of the coalitional configurations in the CCS shall be approached, and therefore there cannot be any polynomial method to find Pareto-optimality. An important advantage of our algorithm is that the average expected payoff of the agents is an increasing function of the time and effort spent by the agents performing the CNA steps. Therefore, if cooperation is beneficial for the agents, using the CNA will always improve their payoffs.

The model we present is not restricted to the super-additive environment, an environment for which there are already several coalition formation algorithms in DAI [12, 16]. However, the generality of the model does not make it inapplicable. As opposed to the majority of solution concepts presented in game theory, we present a detailed method for how the individual agent should act in order to form coalitions that increase its personal payoff.

References

1. R. J. Aumann. The core of a cooperative game without side-payments. *Transactions of the American Mathematical Society*, 98:539–552, 1961.
2. R. J. Aumann, B. Peleg, and P. Rabinowitz. A method for computing the kernel of n-person games. *Mathematics of Computation*, 19:531–551, 1965.
3. M. Davis and M. Maschler. The kernel of a cooperative game. *Naval research Logistics Quarterly*, 12:223–259, 1965.
4. Thomas Dean and Mark Boddy. An analysis of time-dependent planning. In *Proceedings, AAAI88*, pages 49–54, St. Paul, Minnesota, 1988.
5. E. Durfee. What your computer really needs to know, you learned in kindergarten. In *Proc. of AAAI-92*, pages 858–864, California, 1992.
6. S. Kraus and J. Wilkenfeld. Negotiations over time in a multi agent environment: Preliminary report. In *Proc. of IJCAI-91*, pages 56–61, Australia, 1991.
7. S. Kraus and J. Wilkenfeld. A strategic negotiations model with applications to an international crisis. *IEEE Transaction on Systems Man and Cybernetics*, 23(1):313—323, 1993.
8. R. D. Luce and H. Raiffa. *Games and Decisions*. John Wiley and Sons, Inc, 1957.
9. B. Peleg. The kernel of m-quota games. *Canadian Journal of Mathematics of Computation*, 17:239–244, 1965.
10. B. Peleg. *Game Theoretic Analysis of Voting in Commities*. Cambridge University Press, 1984.
11. A. Rapoport. *N-Person Game Theory*. University of Michigan, 1970.
12. O. Shechory and S. Kraus. Coalition formation among autonomous agents: Strategies and complexity. In *Proc. of MAAMAW-93*, Neuchâtel, 1993.
13. R. E. Stearns. Convergent transfer schemes for n-person games. *Transactions of the American Mathematical Society*, 134:449–459, 1968.
14. E. Werner. Toward a theory of communication and cooperation for multiagent planning. In *Proc. of the Second Conference on Theoretical Aspects of Reasoning about Knowledge*, pages 129–143, Pacific Grove, California, March 1988.
15. L. S. Wu. A dynamic theory for the class of games with nonempty cores. *Siam Journal of Applied Mathematics*, 32:328–338, 1977.
16. G. Zlotkin and J. S. Rosenschein. Coalition, cryptography, and stability: Mechanisms for coalition formation in task oriented domains. In *Proc. of AAAI94*, pages 432–437, Seattle, Washington, 1994.

A Knowledge-Level Model of Co-ordination

Sascha Ossowski and Ana García-Serrano

Department of Artificial Intelligence
Technical University of Madrid
Campus de Montegancedo s/n
28660 Boadilla del Monte
Madrid, Spain
Tel: (+34-1) 336-7390; Fax: (+34-1) 352-4819
{ossowski, agarcia}@dia.fi.upm.es

Abstract. Co-ordination is one of the central research issues in Distributed Artificial Intelligence. Most of the efforts in designing co-ordination mechanisms set out from an agent-centred point of view: they see the individual actor with its local beliefs and reasoning capabilities as the foundation of all system properties. Our thesis is that co-ordination is best modelled from an environment-centred, social perspective, that considers the whole, situated system as the starting point of any analysis. We argue that this stance is reflected in Clancey's modification of the Knowledge-level hypothesis. On the basis of this modified hypothesis we introduce a distinction between co-ordination at the Knowledge- and at the Symbol-level. Finally, we present a Knowledge-level model of co-ordination and illustrate it by specifying the co-ordination processes of an intelligent traffic management system.

1 Introduction

The phenomenon of co-ordination is being tackled by a variety of scientific disciplines. Sociologists, by means of observation, want to explain how particular co-ordination mechanisms work within a certain society of people and why they emerged. Economists study the market as a particular co-ordination mechanism. Organisational Theorists not only try to explain the co-ordination behaviour of an organisation, but also aim to predict its future behaviour, assuming the validity of a certain co-ordination mechanism [7].

Within Computer Science, the area of Distributed Artificial Intelligence (DAI) is deeply concerned with the problem of co-ordination. Its focus is different from those described above, as it aims to *design* co-ordination mechanisms for groups of artificial, (boundedly) rational agents. Following the habits of traditional Artificial Intelligence, which was moved by the desire to model individual intelligence, DAI has put the individual agent in the centre of analysis of distributed intelligent systems. Although the boundaries of research agendas in DAI have become blurred, two major paths, represented by the Multi-agent Systems (MAS) and the Distributed Problem Solving (DPS) communities, can be distinguished [9].

---

* This work was supported by the Human Capital and Mobility Program (HCM) of the European Union, contract ERBCHBICT941611

The MAS community within DAI focuses on the development of Agent-Technology. Formal theories are developed to specify requirements for agent behaviour, and agent architectures constructed to operationalise it. Following Dennet's philosophical stance, agents are usually seen as *intentional systems* and described by "mental" terminology [23]. This leads to the so-called BDI-architectures (belief, desire, intention): all agent actions originate from a current set of beliefs (knowledge) and desires (goals), which are linked together by intentions. Co-ordination here is an emergent phenomenon: agents generate individual *commitments* on the basis of their local beliefs and communicate them to other agents, which then change their beliefs, due to their reliance on certain actions to be performed by the former. Gasser [11] notes, that there is a circularity in this approach to model co-ordination: facts and beliefs provide commitments, but they are themselves commitments.

Contrary to MAS, which in general makes assumptions about the properties of individuals and then considers what properties will emerge internally from among the agents, the DPS community deals with macro-phenomena more directly. DPS systems also assume some basic internal properties, such as benevolent agents and common goals, but are primarily concerned with how to achieve desired external properties (robustness etc.) in line with these assumptions [9]. Agent co-ordination strategies are developed, and their effectiveness measured in terms of the degree to which they achieve the required external properties. Co-ordination in DPS architectures is usually concerned with reducing local control uncertainty, which results from partial and possibly conflicting views of the agents of the global system state. Thus, in DPS systems co-ordination is not driven by external, environmental uncertainty, but depends to a greater extent on the internal uncertainty, which is due to the (designed-in) distribution of control.

2 Co-ordination from a Social Perspective

Within the traditional DAI perspective described above both communities try to explain co-ordination behaviour in terms of the attributes of individual agents. Such an agent-centred view distracts from the simple truth, that co-ordination primarily is a social phenomenon: it is a global property of a system and should be characterised in terms that abstract from the properties of the system's components (i.e. its constituent agents). In Sociology and related sciences, such a social perspective on co-ordination, and on the nature of intelligent behaviour in general, is not new. For instance, by the early years of this century Mead had already stated:

> *We are not, in social psychology, building up the behaviour of the social group in terms of the behaviour of the separate individuals composing it; rather we are starting with a given social whole of complex group activities, into which we analyse as elements the behaviour of each of the separate individuals composing it. [...]. For social psychology, the whole (society) is prior to the part (the individual)* [10]

This viewpoint leads to a characterisation of co-ordination in terms of the relations that exist between an entire system and its environment. It stresses the simple fact, that all real systems are *situated*. The behaviour of an intelligent system can only be understood by focusing on its interactions with the real world in a particular situation in which it has to act. Consequently, a model of co-ordination must set out by relating a system's co-ordination behaviour to the characteristics of its environment.

As a prerequisite for this enterprise, the precise meaning of the term under study needs to be further clarified. For this purpose, we adopt the definition of Malone and colleagues [13], who identify co-ordination with *the act of managing dependencies between activities*. This stance already includes many aspects of a social perspective on co-ordination.

Firstly, it does not require several actors to be present in order for the phenomenon of co-ordination to arise. Instead, co-ordination is seen as a result of performing several *activities*, which need not but may be carried out by the same actor.[1]

Secondly, it puts emphasis on the "situatedness" of co-ordinated systems, due to the fact that *dependencies* arise as a consequence of activities being performed in the same environment. Activities interact by referring to the same environmental entity. This is often a physical, usually limited resource, which has to be shared by a set of activities (i.e. the processor of a multi-tasking workstation), or which is produced by one activity and "consumed" by another (i.e. the carburettor in a motor assembly process). However, "immaterial" information resources, such as an interface specification or knowledge about possible task decompositions, can also be related to a system's activities, and consequently need to be considered, too.

Finally, the management of the dependencies, as mentioned in the definition, consists of the exhibition of certain co-ordination behaviour in response to them. For this, the environment has to be perceived, dependencies detected and co-ordination behaviour associated. This process suffers from the *uncertainty* which is intrinsic to co-ordination, as the perception of the environment may be ambiguous, a dependency not detected or out of date due to a change in the environment, outcomes of actions unpredictable etc.. This uncertainty is implied by the environment (or by the system's global perception of the environment), but not by the incomplete views of the overall problem-solving state that some components of the system may have.

From a social perspective, co-ordination can only be modelled by dropping the closed-systems assumption and seeing "situatedness" as the essential system property. A system's co-ordination behaviour is primarily determined by its environment, which is the source of dependency and uncertainty,. in relation to which it envisions and performs its actions.

3 Modelling Co-ordination at the Knowledge-level

The Knowledge-level hypothesis has been introduced by Newell [16] as an attempt to clarify the relation between knowledge on the one hand and symbolic representations on the other. It claims that there is a computer systems level lying directly above the Symbol-level (SL). This Knowledge-level (KL) is characterised by knowledge as the medium and the principle of rationality as the law of behaviour. A KL-description of a system consists of the knowledge and the goals, that an observer ascribes to an "agent", in order that it can exhibit an observed behaviour by applying its knowledge according to the principle of rationality. Such a description is radically different from

---

[1] This accounts for the common sense meaning of the term, according to which one sole agent can act in an "uncoordinated" fashion.

traditional viewpoints, in that it constitutes a (subjective) abstraction, made by an observer, and thus completely ignores any considerations concerning architectures or physical structures. In the following we will apply the KL-hypothesis to the phenomenon of co-ordination.

3.1 Co-ordination and the Knowledge-level

Newell's KL-hypothesis has been shown to be useful in many areas of Artificial Intelligence, such as knowledge acquisition, knowledge-based systems construction etc. [17]. Dietterich first applied it to machine learning. He distinguishes Knowledge-level learning, where a system acquires a new KL description, from Symbol-level learning, where the system learns only to evoke the knowledge it already has faster and more reliable [8]. Schreiber and colleagues transfer the KL-hypothesis to control [19]. In the term Knowledge-level control they include strategic knowledge in regard to task decompositions and orderings as well as meta-knowledge about a system's components. Symbol-level control is concerned with control issues that arise when a particular representation or AI technique is selected to realise a problem-solving component.

A similar distinction concerning co-ordination is complicated by the fact that Newell defines the observed entity, an agent, only vaguely as composed of *a set of actions, a set of goals and a body* [16]. Hence, according to Newell, the term agent refers to a broad variety of things, from a simple computational process up to a complex MAS, the important point is that actions, goals and knowledge can be ascribed to it.[2] This raises the question, as to whether the agents in DAI are the same as the agents that Newell refers to. We claim that, seen from a social stance, this is not the case. This idea is underpinned by Clancey's modification of the KL hypothesis:

> *A KL description is about a situated system, not an agent in isolation. That is, the systems level being described is above that one of individual agents. Therefore, a knowledge-level description cannot be identified with (isomorphically mapped to) something pre-existing inside an individual head, but rather concerns patterns that emerge in interactions the agent has in some (social) world* [2]

Thus, we conceive a KL description as a characterisation of an entire system as a whole, while the system's components, the agents, are SL entities.[3] Consequently, we can distinguish between two levels of co-ordination: Knowledge-level co-ordination refers to whole systems. It is a property, that can be observed by the actions that a certain system performs, in relation to dependencies and uncertainty that exist in its environment. Symbol-level co-ordination results from a specific system architecture, i.e. from the symbolic representation of the system, its internal structure. Hence, the different co-ordination techniques applied in DPS to reduce a system's internal control uncertainty, including individual commitments, exchange of meta-information etc., are located at the SL.

---

[2] In fact, Shoham [20] notes that everything can be described using this kind of mental terminology. The point lies in whether it is advantageous or not to do so.

[3] As van de Velde [22] argues, in practice a KL-description usually imposes a structure on the system's knowledge. This might be interpreted as a set of functional KL-agents. However, these functional "agents" are the result of an observer's description. Thus, they have no a priori characteristics and definitely no physical base at all.

3.2 A Knowledge-level Model of Co-ordinated Systems

The above characterisation provides the basis for a KL-model of co-ordination. The model will be expressed using the KSM methodology [14,15] as a KL modelling language. In this methodology the central modelling primitive is an entity called *Knowledge Unit (KU)*, which represents a knowledge area of the system. A KU is defined by what it knows, described by a collection of knowledge (sub-)areas, and what it is capable of, specified by a collection of *tasks*. Optional Conceptual Vocabularies (CV) provide the conceptual framework, a common ontology, for the knowledge areas to be used in the KU. In the design process the entire knowledge of the system is structured in knowledge areas, which leads to an organisation of the system knowledge in a network of KUs. Dynamic information is contained in information models, called *stores*, which are modified by the accomplishment of tasks. The relationship between tasks and stores is specified by a task-store diagram.

As shown in the KU-net of figure 1, we propose to model a co-ordinated system at the KL by means of a *System KU*, whose knowledge is made up of three areas: world interaction, co-ordination and management knowledge. The system's dynamics are realised by the *operate*-task, which, on the basis of the actual contents of the information models, determines which of the tasks associated to one of these knowledge areas to accomplish next.

The *World Interaction KU* integrates all the knowledge areas necessary for the system's interactions with the environment. These include mere perceiving activities, that build or update the system's model of the environment, as well as environment-manipulating actions, which are intended to change the state of the outside world in a desired way. The latter can be either problem-solving actions, such as the production of a resource, or co-ordination actions, such as the adaptation or standardisation of a resource involved in a producer-consumer relationship.

The *Co-ordination KU* provides a set of co-ordination tasks, that allow for different styles of co-ordination. It covers two knowledge areas. The first one, integrated in the *Structure Discovery KU*, builds a structured abstraction of the system's perception. This is done by *detect*-tasks that use *Detection KUs* to discover dependencies. The *Dependency CV* ontology specifies the taxonomy of dependencies, on the basis of which *detect*-tasks are accomplished.

It is worthwhile pointing out, that the process of structure discovery is far from being trivial. Consider, for instance, the detection of a new task-subtask dependency. Such a fact implies the dynamic generation of a new problem decomposition. So, the structure of a system's problem-solving process is not designed-in a priori, but emerges during problem-solving.

The *Behaviour Association KU* determines which co-ordination behaviour will be shown by the system in response to the detected dependencies. The corresponding association knowledge may be either complicated inference processes, or a simple Knowledge Base, that links every element of the dependency taxonomy to at least one element of the taxonomy of behaviours defined in the *Co-ordination Behaviour CV*. These behaviours constitute co-ordination activities which may have mere internal effects, for instance restricting the envisioned execution time of certain external actions, or may themselves be external, as for the action of adaptation or standardisation of a resource mentioned above.

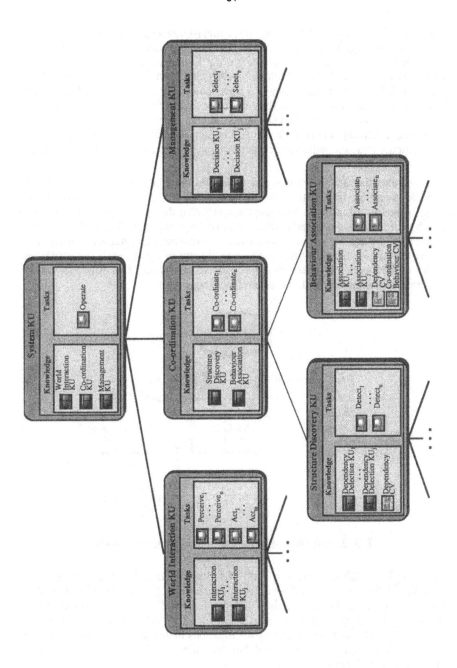

Fig. 1. KU Network of a Co-ordinated System

Finally, the *Management KU* incorporates meta-knowledge on how decisions are taken in the system. In the model, these decisions consider mainly two questions. Firstly, it has to be determined how far in the future the system's actions will be planned, that is, when the *World Interaction KU* and when the *Co-ordination KU* will be activated. Secondly, as there may be various co-ordination behaviours that could be shown in response to one relationship, the most appropriate has to be selected. The content of the *Decision KUs* depends largely on the internal structure of the system. For instance, if the system is made up of an organisation of co-operating agents, their organisational structure will imply certain responsibilities as well as decision and negotiation protocols.

The task-store diagram shown in figure 2 clarifies the relationship between the tasks of the KUs and the information models they create and update. The environmental model integrates all information about the world that the system has accumulated so far. Thus, it is created and updated by the *perceive*-tasks as well as by the expected outcomes of the *act*-tasks. As a consequence of erroneous perceptions or unsuccessful actions the environmental model, as the embodiment of the system's subjective perception of the world, may differ from the objective world state. This difference reflects the (external) uncertainty that the system has to cope with.

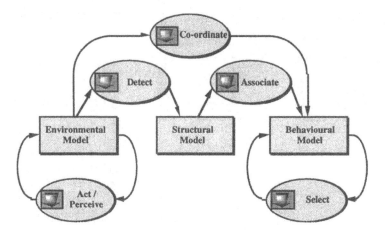

Fig. 2. Task-Store-Diagram of a Co-ordinated System

The *co-ordinate*-task manages two kinds of sub-tasks. First, *detect*-tasks generate the structural model, which constitutes a classification of the perceived environmental entities in terms of the chosen taxonomy of dependencies. Then, *associate*-tasks provide the behavioural model, which contains a set of possible action sequences of the system together with their corresponding justifications in terms of dependencies. *Select*-tasks may decide upon alternatives and thus determine which external actions are actually executed by the system.

In order to model an entire system at the KL using this model, the KU-net shown in figure 1 has to be completed according to the system's special action and perception capabilities, its strategies of decision-making, and its co-ordination abilities. Its interactions with the world will then be determined by the *content* of the knowledge assigned to the KUs.

3.3 Co-ordinating an Intelligent Traffic Management System

We will illustrate the above model by applying it to the problem of co-ordinating the actions of an intelligent traffic management system. For this purpose, we structure the knowledge necessary for this domain according to the above KL-model and describe how its rational application affects the information models (environmental, structural and behavioural), leading to co-ordinated system activities. The concepts used in our example are borrowed from existing real-world intelligent traffic management systems [5,6]. However, for the sake of clarity and to illustrate the co-ordination model better, most concepts are highly simplified.

A traffic management system guides the traffic flow in a certain problem area, say the motorways and ring-roads in eastern Madrid, with the objective of avoiding and/or clearing up congestions. Traffic information is provided to the system by loop detectors, that measure traffic speed, flow and occupancy. It acts by displaying different messages on a set of Variable Message Panels (VMS). A logically coherent set of messages, called a *proposal*, is intended to influence the development of traffic load in the respective zones.

Proposals for different sub-areas are interdependent, thus need to be co-ordinated. Figure 3 shows a simple taxonomy of possible dependencies. Most important is the case of harmful interactions, in that one proposal hinders or even cancels the effectiveness of another. The latter is the case of physical dependencies, when two proposals include different messages to be displayed on the same VMS. Logical dependencies can be classified in two groups: incompatible effects arise within a set of proposals, when two neighbouring panels are envisioned to display contradictory messages. Undesirable effects are present, when one proposal influences the traffic in a way that aggravates another traffic problem, and thus hinders the proposal associated to that problem from being effective. Favourable dependencies exist in the opposite case, when one proposal lightens the traffic problem that gave rise to another.

Figure 3 also depicts possible co-ordination behaviours for managing these dependencies. Congestion warnings for the areas A and B, included in different proposals and to be displayed on the same VMS, can be merged into a "Congestion at A and B"-message on that panel. An alternative proposal for a problem can be selected if the one under consideration is in conflict with another. Finally, it might be possible to adapt interdependent proposals by displaying "harmonising" messages on unused panels.

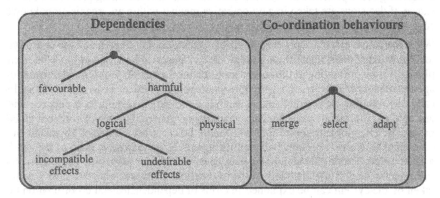

Fig. 3. Taxonomies for Traffic Management

Just like in figure 1, the intelligent traffic management system is modelled by a top-level *System KU*, that integrates three knowledge sub-areas. Dependencies and co-ordination behaviours shown in figure 3 constitute the ontologies included in the *Dependency CV* and *Co-ordination Behaviour CV* Conceptual Vocabularies of our model of co-ordination.

The *World Interaction KU* is endowed with knowledge about two kinds of activities. On the one hand it knows how to perform the environment-changing actions of displaying messages, on the other hand it is capable of perceiving (part of) the environment. We understand perception here in the broad sense of translating sensed numerical data into abstracted symbolic representations. As a consequence, the perception tasks themselves make use of several subordinated knowledge areas: the reception of temporal series of magnitudes such as traffic speed, flow and occupancy from each sensor; the completion of possibly erroneous data from the sensors; the calculation of temporal and spatial gradients etc.; the generation of short-term predictions; and finally the association of prototypic traffic (problem) situations. Both, environment-changing and perception actions construct and update the system's environmental model by modifying its beliefs about the actual traffic situations in different zones.

As a sub-area of the *Co-ordination KU*, the *Structure Discovery KU* provides *detect*-tasks, which access expert knowledge about how drivers choose routes and how panel messages affect this choice. On this basis they associate proposals to the problematic traffic situations of the environmental model and detect dependencies between them. Applying all this knowledge in a rational manner, the *Structure Discovery KU* is capable of generating the structural model. It consists of proposals and dependencies between them, expressed in terms of the taxonomy of dependencies shown in figure 3.

The *Behaviour Association KU*, the other knowledge sub-area of the *Co-ordination KU*, is provided with knowledge on how to associate co-ordination behaviour in response to dependencies. In the case of the *select*-behaviour, its application has only internal effects, as just one of a set of possible proposals is selected. Still, *merge-* and *adapt*-behaviours do have external effects, namely the generation of new messages or the use of new (idle) VMS. The result is a set of globally coherent proposals, similar to a non-linear plan, which constitutes the behavioural model.

Finally, one proposal must be selected for execution. The knowledge necessary for this decision is incorporated in the *Management KU*. It may simply be knowledge concerning static priorities or describe more sophisticated reasoning processes. For instance, if decisions are taken in a co-operative manner within an organisation, knowledge about voting and negotiation procedures may be included.

When the above model is specified at the SL, agents are introduced and the entire system knowledge distributed among them. One example of how this might be accomplished is the following multi-agent architecture: a set of problem identification agents detect problems in some (possibly overlapping) problem areas and associate locally coherent proposals with them. A central co-ordination agent detects dependencies and applies co-ordination mechanisms accordingly, generating a set of coherent global proposals. The global proposal that will finally be executed is selected by means of a co-operative decision process between the agents. If the objective state of the world and the agents' environmental model drift apart, this will be perceived via sensors, the information models updated and envisioned action sequences adapted. Every action taken by the system can be explained in terms of dependencies and co-ordination behaviours. So, we believe that this environment-centred, social perspective on co-ordination will help to build systems that are provided with the potential to show reactive, adaptive and explicatory behaviour.

4 Related Work

Aitken and colleagues [1] use Clancey's modified KL hypothesis to account for global properties of distributed knowledge-based systems. They present a KL-characterisation of MAS from a social perspective and argue in favour of understanding the concept of global coherence in these terms. Durfee and Rosenschein [9] sketch attempts to extend the area of DPS to include systems that cope with changing environments (i.e. situated systems). Jennings [12] presents a sharply contrasting approach, that tries to establish a "co-operation level" on top of the KL[4], in order to account for system properties, such as co-ordination, while describing individual agents at the KL. However, Conte and Castelfranchi [3] conciliate the above positions to a certain degree when they remember the dialectic character that the problem appears to have in real societies: collective (system) properties influence individual (agent) behaviour and vice versa.

Historically, theories of co-ordination, expressed in terms of generic dependencies and co-ordination behaviours, have been restricted to Organisational Research and related sciences. Ultimately, there have been attempts to develop such universal taxonomies in Co-ordination Science and Artificial Intelligence. Malone [13] outlines an abstract, informal framework of dependencies between activities and vaguely associates classes of behaviours to them. Crowston [4] presents a more detailed taxonomy of dependencies between tasks and resources. However, he only depicts examples of related co-ordination behaviour. Van Martial [21] classifies relationships between actions of individual plans in a multi-agent environment. The detection of these dependencies and the association of co-ordination actions is integrated in an operational planning framework. In Decker's GPGP approach [7], agents with only local views of the overall system problem-solving state, detect environmental dependencies, classified in terms of the TÆMS modelling language, and co-ordinate their plans. This is probably the most general operational approach to co-ordination that puts its focal point on environmental aspects. Nevertheless, we believe that problems resulting from co-ordination are not explicitly separated from problems caused by distribution, which does not contribute to conceptual clarity. However, no attempt is known to us that tries to unify these approaches in a common KL-model of co-ordination.

5 Discussion

In this paper we argued in favour of a more social, situated conception of co-ordination. This stance allowed us to construct a KL-model of co-ordination, and thereby shed some light on the relationship between co-ordinated and multi-agent systems on one hand and the Knowledge-level hypothesis on the other. We claimed that co-ordination and distribution of control each address essentially different problems, namely the management of (external) environmental uncertainty contrasted with the management of (internal) control uncertainty.

---

[4] Ultimately, Jennings seems to term this concept "social level".

We aim to unify the above model of co-ordination with the idea of knowledge structuring by means of Knowledge Units at the KL. Different techniques will be developed that provide context-sensitive mappings from KL description of co-ordinated systems to societies of (SL) agents. By augmenting KSM with principled models of co-operation and co-ordination we intend to deepen its support for the construction of (multi-agent) knowledge systems, from their KL-specification down to its actual implementation. We hope that such a tool will contribute to make knowledge systems more robust and less brittle.

References

1. S. Aitken, F. Schmalhofer, N. Shadbolt. "A Knowledge Level Characterisation of Multi-Agent Systems". Wooldridge, Jennings (Eds.): *Intelligent Agents*, Springer, 1995, p. 179-190
2. W. Clancey. "The Frame of Reference Problem in the Design of Intelligent Machines". Vanlehn (Ed.): *Architectures for Intelligence*, Lawrence Erlbaum, 1991, p. 357-423
3. R. Conte, C. Castelfranchi. *Cognitive and Social Action*, UCL Press, 1995
4. K. Crowston. "A Taxonomy of Organisational Dependencies and Co-ordination Processes". *MIT Centre of Co-ordination Science Working Paper No. 174,* 1994
5. J. Cuena, J. Hernández, M. Molina. "Case Presentation of the Use of Knowledge Based Models for Traffic Management - Madrid". *Proc. World Congress on Applications of Transport Telematics and Intelligent Vehicle-Highways Systems,* 1994
6. J. Cuena, J. Hernández, M. Molina. "Using Knowledge-based Models for Adaptive Traffic Management Systems". To appear in: *Transportation Research,* 1995
7. K. Decker. "Environment Centred Analysis and Design of Co-ordination Mechanisms". Ph.D. Thesis. *UMass Computer Science Technical Report 95-XX,* 1995
8. T. Dietterich. "Learning at the Knowledge Level". *Machine Learning 1,* 1986, p. 287-316
9. E. Durfee, J. Rosenschein. "Distributed Problem Solving and Multi-Agent Systems: Comparisons and Examples". *Proc. Thirteenth International Distributed Artificial Intelligence Workshop,* 1994, p. 94-194
10. L. Gasser. "Social conceptions of knowledge and action: DAI foundations and open systems semantics". *Artificial Intelligence 47,* 1991, p. 107-138
11. L. Gasser. "Architectures and Environments for AI". Doc. Fifth Advanced Course in *Artificial Intelligence ACAI'93,* 1993
12. N. Jennings. "Towards a Cooperative Knowledge Level For Collaborative Problem Solving". *Proc. ECAI'92,* 1992, p. 225-228
13. T. Malone, K. Crowston. "The Interdisciplinary Study of Co-ordination". *ACM Computing Surveys 26,* 1994, p. 87-119
14. M. Molina. *Desarrollo de Aplicaciones a Nivel Cognitivo Mediante Entornos de Conocimiento Estructurado.* Ph.D. Thesis. Technical University of Madrid, School of Computer Science, 1993
15. M. Molina, J. Cuena. "Knowledge Oriented and Object Oriented Design: the Experience of KSM". *Proc. Knowledge Acquisition Workshop KAW-95,* 1995
16. A. Newell. "The Knowledge Level". *Artificial Intelligence 18,* 1982, p. 87-127
17. A. Newell. "Reflections on the Knowledge Level". *Artificial Intelligence 59,* 1993, p. 31-38
18. T. Pylyshyn. *Computación y Conocimiento.* Editorial Debate, 1988
19. G. Schreiber, H. Akkermans, B. Wielinga. "On Problems with the Knowledge Level Perspective". *Proc. Knowledge Acquisition for Knowledge-Based Systems Workshop KAW-90,* 1990

20. Y. Shoham. "Agent-oriented programming". *Artificial Intelligence 60*, 1993, p. 51-92

21. F. van Martial. *Co-ordinating Plans of Autonomous Agents*, Springer, 1992

22. W. van de Velde. "Issues in Knowledge Level Modelling". David, Krivine, Simmons (Eds.): *Second Generation Expert Systems*, Springer, 1993, p. 211-231

23. M. Wooldridge, N. Jennings. "Intelligent Agents: Theory and Practice". *Knowledge Engineering Review 10*, 1995, p. 115-152

An Evolutionary Model of Multi-Agent Systems

Bengt Carlsson
Department of Computer Science and Business Administration
University of Karlskrona/Ronneby
Ronneby, Sweden

Lund University Cognitive Science
Kungshuset, Lundagård
S-22350 Lund,Sweden
bengt.carlsson@pt.hk-r.se
Tel: +46 (0)45780060
Fax: +46 (0)45780006

Abstract

A multi-agent system can be viewed as an evolutionary system where each individual (agent) acts on a basis of population dynamics and stability under the criterion of Darwinian fitness. This "survival of the fittest" or "best for individual (or gene or agent)" argument can be used to find evolutionary stable strategies within the multi-agent system. The aim of this paper is to introduce some of these evolutionary ideas to the multi-agent society.

The differences between a product maximising mechanism (PMM) and an evolutionary stable strategy (ESS) are discussed. In a PMM the utility for an agent, in a mixed joint plan, is the positive difference between the maximum expected cost that the agent is willing to pay in order to achieve his goal, and his expected part of the outcome. In an ESS the fitness of the utility function does not have to be positive because zero fitness is a state which can be both improved and weakened. Both PMM and ESS are, according to game theory, individual rational and pareto optimal but they address different kinds of problems. If we know the maximum expected costs that each agent is willing to pay to achieve his goal, it is possible to use a PMM. If we instead know the agent's zero fitness, which is assumed to be the same for all agents, it is possible to use an ESS.

As an example, a solution to the telephone call competition among two agents with an overbid-underbid strategy (or hawk-dove strategy in the terminology used by evolutionary biologists) is demonstrated by using an ESS. This solution is compared with one in which the best bid wins and is paid the second price.

In an extended asymmetric game between two competing agents, one mixed and one pure evolutionary stable strategy are found. It is proposed that both these strategies will bring the competing agents near the "real" value and accordingly can be alternatives to the best bid wins, get second price idea.

Keywords: Multi-Agent System (MAS), Evolutionary Stable Strategy (ESS), Product Maximising Mechanisms (PMM), Distributed Artificial Intelligence (DAI), Game Theory, Agent Interaction.

Introduction

In Distributed Artificial Intelligence (DAI), there have been several areas of research that have addressed the problem of multi-agent coordination. One approach taken in both multi-agent systems and evolutionary theory is based on models in game theory. Some central issues for multi-agent coordination involve planning and negotiation. Is it just a question of coordination among agents with no basic conflicting goals, or will agents cooperate only when it is in their own interest to do so? The concept of rationality, that the player will behave rationally and according to some criterion of self interest, can be used to model agent activity in a multi-agent negotiation using a payoff matrix (*Genesereth et al. 1988, Rosenschein and Genesereth 1988*).

Evolutionary biologists look at members of a population as if they are playing games against each other, and model the population dynamics and equilibrium which arises. So, what will happen if we let agents behave like living creatures? Is the behaviour in the interest of the species, in the interest of the whole group of agents, or in the interest of an individual agent? For several decades most evolutionary researchers have used the "best for individual" argument (*Wilson 1975, Dawkins 1976, 1982, Maynard Smith 1989*) to model evolution.

A Telephone Call Competition Example—Some Basic Ideas

I will use a telephone call competition example to introduce an evolutionary model in a multi-agent setting. In the USA (and Sweden) there are several long distance telephone companies competing for customers. At any time a customer can typically have a choice between the companies.

Rosenschein and Zlotkin give an example of a telephone call competition in their paper: 'Designing conventions for automated negotiation' (*Rosenschein and Zlotkin 1994a*). They ask what will happen if a customer places a long-distance call, and a microprocessor within the telephone automatically collects bids from the various carriers? This problem can be described as a task for multi-agent system. Let's look at the auction models given by Rosenschein and Zlotkin and an evolutionary model of negotiation.

Auction models

Best bid wins. The winning carrier receives a pay equal to the amount that was bidden by the carrier. This means that prices can be completely dynamic and that the company, instead of investing in costly advertising campaigns will profit by lowering their costs instead. But what will happen if the company instead invests in strategic reasoning? Perhaps it is possible to get a larger profit by figuring out the strategies of different competitors. Consequently, you may not have to give your lowest possible bid in order to get the contract.

Best bid wins and gets paid second price (Vickrey's Mechanism). The carrier that wins gets paid a price equal to the second lowest bid. Rosenschein and Zlotkin point out that this protocol will force the company not to underbid or overbid because its own bid will never affect how much money it gets. "We have got a distributed, symmetric, stable, simple and efficient mechanism for these self-motivated machines to use" *(Rosenschein and Zlotkin 1994a)*.

An Evolutionary Model

The evolutionary model is primarily based on what is known as "Nash's Extension of Bargaining Problem" *(Luce and Raiffa 1957)*. Classic bargaining theory is focused on the prediction of outcomes, and on certain assumptions concerning the players and the outcomes themselves. A fair solution predicts an agreement among players that will maximise the product of the players' utility under the assumption that the deal will be individually rational and pareto optimal. *(Nash 1950, 1953)* An agent would at least get the same as he would otherwise without an agreement. There does not exist another deal that dominates the current deal.

The evolutionary model will focus on two assertions. Firstly Rosenschein and Zlotkins' Product Maximising Mechanisms (PMM), "Since we assume that the agents' designers are basically interested in their own goals, we want to find interaction techniques that are 'stable'.." *(Rosenschein and Zlotkin 1994b p. 5)*. Secondly Maynard Smith and Price's theory *(Maynard Smith and Price 1973)* of evolutionary stable strategies (ESS), where "..the criterion of rationality is replaced by that of population dynamics and stability and the criterion of self-interest by Darwinian fitness." *(Maynard Smith 1982 p. 2)*. ESS have been used in several disciplines outside biology including political science *(Axelrod and Hamilton 1981 Axelrod 1984)* and natural language *(Wärneryd 1995)*.

PMM— Product Maximising Mechanisms

A product maximising mechanism includes both a protocol and a strategy. The entire class of such mechanisms stipulates that the protocol is symmetrically distributed, the strategy is in equilibrium with itself and implies both an individual rationality and pareto optimality of the agreed-upon deals.

A PMM has a utility for an agent that is the difference between the worth of a final state and the cost that an agent has to pay to get to the final state. In other words, if an agent can minimise its stand-alone cost it can perform its single-agent best plan or, if the agent finds itself in a shared environment, fulfill a mixed joint plan. This mixed plan will be the difference between the final state's worth to the agent, and the expected work to which the deal commits the agent.

ESS—Evolutionary Stable Strategies

Assume that players can predict the behaviour of their opponents from their past observations of play in "similar games", either with their current opponents or with "similar" ones. If players observe their opponents' strategies and receive many observations, then each player's expectations about the play of his opponents converges to the probability distribution corresponding to the sample average of play he has observed in the past. In this case, if the system converges to a steady state, the steady state must be a Nash equilibrium *(Fudenberg and Tirole 1991)*, or with the terminology used here an ESS.

This can be seen as a large-population model of adjustment to Nash equilibrium, an adjustment of population fractions by evolution as opposed to learning. An ESS is a strategy in that if all the members of a population adopt it, then no mutant strategy could invade the population under the influence of natural selection *(Maynard Smith and Price 1973)*. Or in other words: an ESS is a strategy which does well against copies of itself. A successful strategy is one which dominates the population, therefore it will tend to meet copies of itself. Conversely, if it is not successful against copies of itself, it will not dominate the population.

An ESS has a utility for an agent that is its fitness function. A contest between agents can improve, reduce or have zero fitness. Observe that zero fitness means that the fitness of an agent does not alter as a result of a contest with other agents.

A Symmetric Example of Telephone Call Competition

The Hawk-Dove Game

One of the simplest examples of an ESS is the hawk-dove game. It can be described as a payoff matrix between a hawk *(H)* and a dove *(D)* (given a symmetric pairwise contest) where *P* stands for profit and *C* for cost *(Maynard Smith 1982)*.

	H	D
H	½ (P-C)	P
D	0	½ P

Fig. 1. Hawk-Dove matrix

A hawk escalates and continues until fitness is reduced by a cost *(C)* or until gaining the profit *(P)* . If two hawks meet there is a fifty-fifty chance for each agent to obtain the profit.

A dove retreats at once if opposition escalates. A hawk always obtains the profit against a dove but does not alter the fitness of the dove. This means that the doves either have profits or costs, in a biological setting, like injury or time delay. When two doves meet they will share the profit without having to pay for it. This is originally an example from the biological field although not about real hawks and doves (the doves can indeed really be aggresive). Instead it is about different behaviours among animals or, as I suggest in this paper, among agents. I will use Rosenschein and Zlotkin's negotiation mechanism for State Oriented Domains *(Rosenschein and Zlotkin 1994b p. 97 - 100)* and Maynard Smith's basic model for the Hawk-Dove game *(Maynard Smith 1982 p. 12 - 17)* to describe these agents.

Given an infinite population of agents in a State Oriented Domain < **S, A, J, f** > where

- **S** is the set of all possible world states

- **A** = { A_1, A_2, ...,A_n} is an ordered list of agents.

- **J** is the set of all possible strategies.

- **f** is a function f: **J** → $\mathbf{R}^n$ where **J** = (J_1, J_2, ...,J_n)

Let's assume that *I* is a stable strategy and *M* some mutant strategy with a small frequency *p* of *M* strategists. The stable strategy must have evolved through population changes in time. If almost all members of the population adopt I, then the fitness of these typical members is greater than that of any possible mutant. If, on the other hand, M's frequency p is large, it could invade the population and I would not be stable.

Let $\sigma = (J{:}p)$ be a mixed joint plan where $J \in \mathbf{J}$. J_i is a strategy for A_i and in this case there are two strategies $I = (I_1, I_2)$ and $M = (M_1, M_2)$. Consider a population consisting mainly of I, with a small frequency p of some mutant strategy M. Before the contest all individuals have a fitness $Fitness_0(\sigma)$, a zero fitness which can be both improved and reduced.

$$Fitness_I(\sigma) = Fitness_0(\sigma) + (1\text{-}p)\ \mathbf{f}(I_1,I_2)_I + p\ \mathbf{f}(I_1,M_2)_M$$

$$Fitness_M(\sigma) = Fitness_0(\sigma) + (1\text{-}p)\ \mathbf{f}(M_1,I_1)_I + p\ \mathbf{f}(M_1,M_2)_M$$

Since I is stable it must hold that $Fitness_I(\sigma) > Fitness_M(\sigma)$ and this requires, since $p \ll 1$, for all $M \neq I$:

either $\mathbf{f}(I_1,I_2)_I > \mathbf{f}(M_1,I_1)_I$

or $\mathbf{f}(I_1,I_2)_I = \mathbf{f}(M_1,I_1)_I$ and $\mathbf{f}(I_1,M_2)_M > \mathbf{f}(M_1,M_2)_M$

Any strategy satisfying this condition is an evolutionary stable strategy, ESS.

In the hawk-dove contest D is not an ESS, because $\mathbf{f}(D_1,D_2)_D < \mathbf{f}(H_1,D_1)_D$ ($\frac{1}{2}P < P$) . A population of doves can be invaded by a hawk mutant.

Hawk is an ESS if $\mathbf{f}(H_1,H_2)_H > \mathbf{f}(D_1,H_1)_H$ or as in Fig. 1 $\frac{1}{2}(P\text{-}C) > 0$ which means $P > C$. If it is worth risking injury to obtain the resource, H is the only sensible strategy.

Neither H nor D is an ESS if $P < C$, instead there will be a mixed strategy. If there is a value Q which makes I an ESS of the Hawk-Dove game, we can find it by solving the equation

$\mathbf{f}(H)_I = \mathbf{f}(D)_I$

Therefore

$$Q\ \mathbf{f}(H_1,H_2)_H + (1 - Q)\ \mathbf{f}(H_1,D_2)_D = Q\ \mathbf{f}(D_1,H_1)_H + (1 - Q)\ \mathbf{f}(D_1,D_2)_D$$

or from Fig. 1:

$$\tfrac{1}{2}(P - C)\ Q + P\ (1 - Q) = \tfrac{1}{2}P\ (1 - Q) \text{ or } Q = P/C$$

is an evolutionary stable strategy, ESS.

An Underbid-Overbid Example of Telephone Call Competition

In the next section I will use an evolutionary stable strategy (ESS) and two new behaviours, underbid and overbid as an alternative to the earlier mentioned Vickrey's mechanism. This is a Hawk-Dove strategy where the underbid is a hawk and the overbid a dove.

Underbid (U): Will always make an underbid. If U meets another U he will either lose or win. If U meets an O he will always win and gain all the profit.

Overbid (O): Will always make an overbid. *O* will always lose against a *U*. If *O* meets another *O* they will (statistically) share the profit, *P*.

Let's look again at a payoff matrix where *P* stands for profit, *C* for cost and *R* for real value.

	U	*O*
U	½ (P-C)	P
O	0	½ P

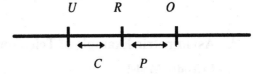

Fig. 2. Underbid-Overbid matrix

Overbid is not an ESS because $\frac{1}{2}P < P$; a population of overbids can be invaded by an underbid mutant.

Underbid is an ESS if $\frac{1}{2}(P - C) > 0$ or $P > C$. In other words, if it is worth risking an underbid to obtain the result, *U* is the only sensible strategy.

But what if $P < C$? Neither *U* nor *O* is an ESS. Instead there is a mixed ESS because $\frac{1}{2}(P - C) < 0$ and $\frac{1}{2}P < P$ and the ESS will adopt *U* with probability

$$p = (P - \tfrac{1}{2}P)/(P + 0 - \tfrac{1}{2}(P - C) - \tfrac{1}{2}P) = \tfrac{1}{2}P/(\tfrac{1}{2}P - \tfrac{1}{2}(P - C)) = P/C$$

and *O* with probability

$$1 - p = 1 - P/C.$$

In other words, there can be both underbids and overbids but the proportions between them will vary according to the different values of *C* and *P*. This will be an aggressive strategy where the underbid company will benefit if it meets an overbid but will eventually be paid under the real value if it meets another underbid. The *underbid-overbid contest* is an ESS with underbid or a mixed strategy (underbid-overbid) as the stable strategy.

Rosenschein and Zlotkin deny that a company, in the telephone call example, should make an underbid or an overbid instead of a "correct" bid. My opinion is that this will be profitable if there are the *right proportions of underbids and overbids in the population.* Let's take one example to illustrate this. The real value for a phone call is 18 cents/minute and the underbid takes 10 cents/minute and the overbid 20 cents/minute, this means that *C = 8* and *P = 2:*

	U	*O*
U	½(2-8)= -3	2
O	0	½•2=1

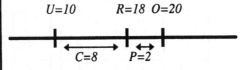

Fig. 3. Example matrix

This is a mixed evolutionary stable strategy with the *proportion p = (2 -1)/(2 - (-3) -1) = 0.25* for the underbid (infinite number of contests, two contestants, no learning). Underbid will get *0.25 • (-3) + 0.75 • 2 = 0.75* and overbid will *get 0.25 • 0 + 0.75 • 1 = 0.75*. On average they will do equally well. This model does not depend on whether each company adopts just one strategy or utilizes overbid strategy during 75 % of the time and underbid during 25 % *(Maynard Smith 1982)*.

An Asymmetric Example of Telephone Call Competition

An Extended Model

There are some constraints in the model above: underbids and overbids all have the same value and there is nothing in the model which may force the underbid or overbid to come near the correct bid. Let's look at an asymmetric game where the value of the bid is different (the first value in the matrix belongs to Agent$_1$ and the second to Agent$_2$). The asymmetric game will have to be repeated an infinite number of times either by different companies or the same company with different bids *(Maynard Smith and Price 1973)*. The example below will be a kind of hawk-dove contest with some restrictions.

U_1 and U_2 will on average share the cost = *min $((P_1-C_1)$; $(P_2-C_2))$/2* for each and O_1 and O_2 will on average share the profit = *min(P_1 ; P_2)/2* for each. In a single contest the lowest bid will win *min $((P_1-C_1)$; $(P_2-C_2))$* or *min(P_1 ; P_2)*.

	U_2	O_2
U_1	*min $((P_1-C_1)$; $(P_2-C_2))$/2* *(min $((P_1-C_1)$; (P_2-C_2)) or 0)*	P_2 /0
O_1	0 / P_1	*min(P_1 ; P_2)/2* *(min(P_1 ; P_2) or 0)*

Fig. 4. Asymmetric game

Pure-strategies are found by testing each cell of the matrix. I have used the matrix below to find a pure-strategy Nash equilibrium *(Fudenberg and Tirole 1991)*. In this case *a,b* and *g,h* respectively have the same value.

	U_2	O_2			U_2	O_2
U_1	*a , b*	*c , d*		U_1	$a \geq e, b \geq d$	$c \geq g, d \geq b$
O_1	*e , f*	*g , h*		O_1	$e \geq a, f \geq h$	$g \geq c, h \geq f$

Fig. 5. Pure-strategy Nash equilibrium

(U_1, U_2) is an ESS if and only if $min\ ((P_1-C_1);\ (P_2-C_2))/2\ \geq\ 0$.
This means $P_1 \geq C_1\ AND\ P_2 \geq C_2$.

(O_1, U_2) is an ESS if and only if $0 \geq min\ ((P_1-C_1);\ (P_2-C_2))/2$
and $P_1 \geq min(P_1;\ P_2)/2$. This means $P_1 \leq C_1\ AND/OR\ P_2 \leq C_2$.

(O_2, U_1) is an ESS if and only if $0 \geq min\ ((P_1-C_1);\ (P_2-C_2))/2$
and $P_2 \geq min(P_1;\ P_2)/2$. This means $P_1 \leq C_1\ AND/OR\ P_2 \leq C_2$.

(O_1, O_2) is not an ESS because $min(P_1;\ P_2)/2 \leq P_2\ (or\ min(P_1;\ P_2)/2 \leq P_1)$.

To determine the *mixed-strategy* equilibrium let p be the probability Agent$_1$ plays U_1 and let q be the probability Agent$_2$ plays U_2.

$$pb + (1-p)f = pd + (1-p)h \text{ and } qa + (1-q)c = qe + (1-q)g$$

$p = (P_1 - min(P_1;\ P_2)/2) / (P_1 - min(P_1;\ P_2)/2 - 0 - min\ ((P_1-C_1);\ (P_2-C_2))/2)$ which means that $min\ ((P_1-C_1);\ (P_2-C_2))/2 < 0$ which means $P_1 < C_1\ AND/OR\ P_2 < C_2$

$q = (P_2 - min(P_1;\ P_2)/2) / (P_2 - min(P_1;\ P_2)/2 - 0 - min\ ((P_1-C_1);\ (P_2-C_2))/2)$ which means that $min\ ((P_1-C_1);\ (P_2-C_2))/2 < 0$ which means $P_1 < C_1\ AND/OR\ P_2 < C_2$.

This is the same as (O_1, U_2) and (O_2, U_1) above so there is one pure strategy (U_1, U_2) with $P_1 \geq C_1\ AND\ P_2 \geq C_2$ and one mixed strategy with $P_1 < C_1\ AND/OR\ P_2 < C_2$.

An Example AT&T vs MCI

Assume that the two companies know that the "real" cost for long distance calls is 18 cents/min (How and why two companies have the same "real" costs is a completely different question). Let's suppose that both companies have one underbid and one overbid strategy. These strategies are supposed to be randomised an infinite number of times. It is also supposed that an underbid strategy will get the price offered by the overbid strategy in a contest between them (hawk-dove contest). When two overbids meet they will share the minimum value and when two underbids meet they will also share their minimum value (not the same). This will in the future be simulated in an environment where the different telephone companies can choose their prices as they wish. In this paper I will use a predefined example (Fig. 6).

	MCI	
AT&T	*Underbid (15 cents)* $C_2 = 3.0$	*Overbid (19 cents)* $P_2 = 1.0$
Underbid (15.5 cents) $C_1 = 2.5$	*Min (2.0-2.5; 1.0-3.0)/2 = -1.0 cents (-2.0 cents or 0 cents)*	*1.0 cents / 0 cents*
Overbid (20 cents) $P_1 = 2.0$	*0 cents / 2.0 cents*	*Min (2.0; 1.0)/2 = 0.5 cents (1.0 cents or 0 cents)*

Fig. 6. AT&T vs MCI

This will be a mixed ESS because P_1 $(2.0) < C_1$ (2.5) (or equally well $P_2 < C_2$) and has the following probabilities:

$p = (0.20 - 0.05) / (0.20 - 0.05 + 0.10) = 3/5$

$q = (0.10 - 0.05) / (0.10 - 0.05 + 0.10) = 1/3$

Let's look at this in more detail:

1. Both companies know the real value which is the same for both, 18.0 cents

2. AT&T gives two bids 15.5 cents and 20.0 cents, and MCI gives two bids 15.0 cents and 19.0 cents. They do not know each others' bids.

3. Probabilities p and q are computed according to Fig. 6. There are four possible outcomes in the matrix with different probabilities.

The customer has the following choices: (Vickrey's Mechanism shows the outcomes of this strategy, with the brackets, [ ], in Fig. 7 indicating a hesitant choice according to a possible underbid).

Probability	3/15	6/15	2/15	4/15
Customers choice	15.0 cents (MCI)	15.5 cents (AT&T)	15.0 cents (MCI)	19.0 cents (MCI)
	15.5 cents (AT&T)	19.0 cents (MCI)	20.0 cents (AT&T)	20.0 cents (AT&T)
Customer choose	15.0 cents (MCI)	15.5 cents (AT&T)	15.0 cents (MCI)	19.0 cents (MCI)
Customer get	16.0 cents (MCI)	19.0 cents (AT&T)	20.0 cents (MCI)	19.0 cents (MCI)
Best bid wins, get second price	[15.5 cents (MCI)]	[19.0 cents (AT&T)]	[20.0 cents (MCI)]	20.0 cents (MCI)

Fig. 7. Different outcomes: AT&T - MCI example

5. AT&T and MCI get a report of the customers' choice but have no knowledge of the strategy of the other company. They can keep or change their strategies.

Discussion

In a product maximising mechanism (PMM), the utility for an agent in a mixed joint plan is the difference between the maximum expected cost that the agent is willing to pay in order to achieve his goal and his expected part of the result. A utility is always greater or equal to zero and the negotiation set is the combination of all mixed joint plans that are both individual rational and pareto optimal.

In an evolutionary stable strategy (ESS), the fitness of the utility function, does not have to be positive because zero fitness is a status quo state which can both be improved and reduced. In an ESS a mixed joint plan, with a stable strategy I and a mutant strategy M, can be described by:

$\text{Fitness}_I(\sigma) = \text{Fitness}_0(\sigma) + (1-p)\, \mathbf{f}(I_1,I_2)_I + p\, \mathbf{f}(I_1,M_2)_M$

$\text{Fitness}_M(\sigma) = \text{Fitness}_0(\sigma) + (1-p)\, \mathbf{f}(M_1,I_1)_I + p\, \mathbf{f}(M_1,M_2)_M$

In a PMM the same mixed joint plan, with a cost function $\mathbf{c}\colon J{\to}(\mathbf{R+})^n$ and a maximum expected cost that agent i is willing to pay $\text{Mcost}_i(\sigma)$, is illustrated by:

$\text{utility}_I(\sigma) = \text{Mcost}_I(\sigma) - (1-p)\, c(I_1,I_2)_I - p\, c(I_1,M_2)_M$

$\text{Utility}_M(\sigma) = \text{Mcost}_M(\sigma) - (1-p)\, c(M_1,I_1)_I - p\, c(M_1,M_2)_M$

If we know the maximum expected costs that each agent is willing to pay to achieve his goal it is possible to use a PMM. If we instead know the agent's zero fitness, which is assumed to be the same for all agents, it is possible to use an ESS.

Telephone call competition between two agents using underbid and overbid is an example of an ESS, which is quite straightforward in the symmetrical case but more complicated in the asymmetrical case. It is not in the scope of this paper to outline all the details, and in fact, it will probably be necessary to simulate the ideas below to see if they can stand a real competition among agents.

Figure 8 below shows the extended Hawk-Dove contest. A black area indicates a single stable point and grey areas the mixed strategy.

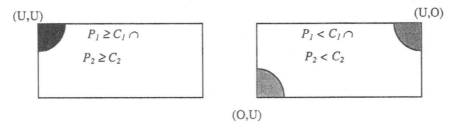

Fig. 8. Types of ESS in extended Hawk-Dove contest

The concept of an ESS is that no different strategies can invade the population.

In a *pure strategy*, it must hold that $P_1 \geq C_1$ and $P_2 \geq C_2$. The Agent with the lowest underbid gets the value *min* $((P_1\text{-}C_1); (P_2\text{-}C_2))/2 \geq 0$. If an Agent wants a large profit it has to give a high overbid and a low underbid, but it is the lowest underbid who takes it all. The contest will bring the underbid near the overbid in the $P_1 \geq C_1, P_2 \geq C_2$ scenario (you can not benefit from giving a moderate underbid because of the risk of losing the contest), and this will bring the price near the real value.

In the *mixed strategy*, it must hold that $P_1 < C_1$ and $P_2 < C_2$. If an Agent wants to minimise its loss, it has to give a C value near the P value which follows that the overall price has to be close to the real value.

The remaining cases $(P_1 > C_1 ; P_2 < C_2$ and $P_1 < C_1 ; P_2 > C_2)$ are not any evolutionary stable strategies, but they can be brought back to either pure or mixed ESS. If Agent$_1$ get a higher payoff where $P_1 > C_1 ; P_2 < C_2$ (or similar for $P_1 < C_1 ; P_2 > C_2$)

this will force Agent$_2$ to let $P_2 > C_2$ and we are back in the pure ESS state. If Agent$_1$ get a weaker payoff this will force Agent$_1$ to let $P_1 < C_1$ and we are back in the mixed ESS.

My conclusion is that the *extended Hawk-Dove contest* will force the agents towards the "real" value. The constraints to do this are weaker than in Vickrey's Mechanism because both the underbid and overbid are allowed. The contest will bring the underbid near the overbid and if P and C are equally far away, they will get the "real" value.

The aim of this paper was to introduce the theory of evolutionary stable strategies (ESS) in a multi-agent world. This means that agents following such a strategy will not be invaded by agents with alternative strategies (but there can be more than one ESS). Such a successful strategy is one which dominates the population of agents, therefore it will tend to meet copies of itself. Our multi-agent society will be a robust society, resisting invasion of agents with alternative strategies. I think that for strictly competitive games the *extended Hawk-Dove contest* is an alternative to the *Vickrey's Mechanism*.

Acknowledgements

I would like to thanks Rune Gustavsson, Peter Gärdenfors and Staffan Hägg for their valuable comments on this paper and Tim Chase for correcting my English.

References

R. Axelrod, *The evolution of co-operation,* New York: Basic Books, 1984.

R. Axelrod and W.D. Hamilton, "The evolution of co-operation", *Science vol. 211,* 1981.

R. Dawkins, *The Selfish Gene,* Oxford University Press, 1976.

R. Dawkins, *The Extended Phenotype,* W. H. Freeman and Company, 1982.

D. Fudenberg and J. Tirole, *Game Theory,* MIT Press, 1991.

M. Genesereth and M. Ginsberg and J. Rosenschein "Co-operation without communication", *in Bond and Gasser Readings in Distributed Artificial Intelligence,* Morgan Kaufmann, 1988.

R. D. Luce and H. Raiffa, *Games and decisions,* Dover Publications Inc, 1957.

J. Maynard Smith and G.R. Price, "The logic of animal conflict", *Nature vol. 246,* 1973.

J. Maynard Smith, *Evolution and the theory of games,* Cambridge University Press, 1982.

J. Maynard Smith, *Evolutionary Genetics,* Oxford University Press, 1989.

J. F. Nash, "The bargaining problem", *Econometrica 28,* 1950.

J. F. Nash, "Two-person cooperative games", *Econometrica 21,* 1953.

J. Rosenschein and M. Genesereth, "Deals among rational agents", *in Readings in Distributed Artificial Intelligence,* Morgan Kaufmann, 1988.

J. Rosenschein and G. Zlotkin, "Designing conventions for automated negotiation", *AI Magazine, 1994a* .

J. Rosenschein and G. Zlotkin, *Rules of Encounter,* MIT Press, 1994b.

E. O. Wilson, *Sociobiology-The abridged edition,* Belknap Press 1980.

K Wärneryd, Language, "Evolution, and the Theory of Games", *in Casti and Karlqvist Co-operation & Conflict in General Evolutionary Processes,* Wiley-Interscience, 1995.

Engagement and Cooperation in Motivated Agent Modelling

Michael Luck[1] and Mark d'Inverno[2]

[1] Department of Computer Science, University of Warwick, Coventry, CV4 7AL, UK.
Email: mikeluck@dcs.warwick.ac.uk
[2] School of Computer Science, University of Westminster, London, W1M 8JS, UK.
Email: dinverm@westminster.ac.uk

Abstract. The title of this paper suggests two distinct aspects of the models that we propose and consider. The first of these is the modelling of other agents by *motivated* agents. That is to say that the act of modelling is itself motivated and constrained by the agent doing that modelling. The second aspect is that all such models will also be of motivated agents. It is not sufficient merely to know what other agents are like, but also to know why they are like that. This *why* aspect is what provides the extra information that allows a greater understanding of the interactions between entities in the world, and consequently provides for more resilient agents capable of effectively dealing with new and unforeseen circumstances in an uncertain world. Previous work has described a formal framework for agency and autonomy in which agents are viewed as objects with *goals*, and autonomous agents are agents with *motivations*. This paper considers the nature of cooperation within that framework. We identify distinct kinds of interaction, depending on the nature of the entities involved. In particular, we describe and specify the differences that arise in these interactions which we characterise as *engagements* of non-autonomous agents, and *cooperation* between autonomous agents.

1 Introduction

Recent work in artificial intelligence has begun to investigate many aspects of single-agent and multi-agent systems. One reason for the concentration of effort into agent-oriented work is that much previous research was concerned with toy problems unrelated to the embodiment of the problem-solver, or whether it was situated in a real external environment. Thus, though significant advances have been made through such work, these have been limited since many of the solutions developed will not adapt well to more realistic scenarios. The relatively recent recognition of these limitations has been a key motivating factor in the research and development of *agents* as systems capable of intelligent behaviour in a resilient and flexible way.

The rapid growth of the field, however, has led to diverse notions of agents and autonomous agents, which are only now, at least partly, becoming reconciled[14]. Our work seeks to contribute to that reconciliation by providing formal but accessible models of agents and autonomous agents, and their interactions.

In previous work, we have proposed definitions of *agency* and *autonomy,* and described how autonomy is distinct but is achieved by *motivating* agency [6] and generating goals [7]. This paper considers aspects of agent modelling within that framework, and describes how the framework provides useful structure that can be exploited by intelligent agents for more effective operation. We begin by outlining previous work on the agent hierarchy. Then we consider the types of interactions that occur in the world, distinguishing in particular between *engagements* of non-autonomous agents and *cooperation* between autonomous agents. We end by illustrating how these relationships structure the information that is available to an agent enabling a more appropriate use of resources.

2 A Framework for Agency and Autonomy

In essence, our framework is a three-tiered hierarchy of entities comprising *objects, agents* and *autonomous agents.* In this hierarchy, all known entities are objects of which some are agents, and of these agents, some are autonomous agents. This is shown as a Venn diagram in Figure 1. Note the addition of some extra categories to be used later in this paper. The central set covers autonomous agents, the ring enclosing it covers non-autonomous or *server-agents* (SAgents in the diagram), and the outer ring covers non-agent objects or *neutral-objects* (NObjects). This section briefly outlines the agent hierarchy. Many details are omitted, and a more complete treatment can be found in [6].

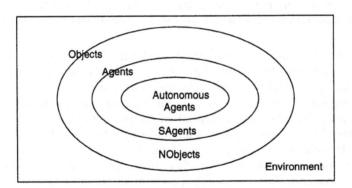

Fig. 1. The Entity Hierarchy

An object is just something with abilities and attributes. For example, the attributes of a table specify that it is stable, wooden, brown and has a flat surface, while its capabilities specify that it can support things. Another well-used example is a robot without a power supply whose capabilities are limited and include just those which rely on its physical presence, such as supporting things, weighing things down, and so on. Its attributes specify that it is blue, that it is upright, that it is large, and so on.

An agent is an object that is (typically) useful to another agent where this usefulness is defined in terms of satisfying that agent's goals. In other words, an agent is an object with an associated set of goals. For example, if I support my computer on a table, then the table is my agent for supporting the computer. The table may not actually *possess* the goal, but is certainly satisfying it. A particular object may produce different instantiations of agents which are created in response to another agent. Agency is thus *transient*, and objects may become agents at some times, while those agents may revert to objects at other times.

For example, if the robot of the earlier example now has a power supply, there are two possibilities. In the first case, the robot does not have a goal and can therefore use its actuators only in a random way, and must be considered an object. In the second case, the robot does have a goal, and this allows it to use its actuators in a *directed* way, such as riveting a panel onto a hull. This means that it is an agent. The goal need not be explicit, but can instead be implicit in the robot's design.

In this sense, an agent can be benevolent. If a robot is designed simply to accept requests to pick up panels and rivet them onto a hull without any evaluation of the request, then it is benevolent. Moreover it is an agent for the *requesting* agent. This is important, for by modelling the relationship of one agent satisfying the goal of another, we can infer much about the situation. For example, if I want the robot to be my agent and satisfy my goals by riveting particular panels for me, then I may have to negotiate with the agent whose goals are currently being satisfied by the robot. Thus agency indicates a relationship between two or more parties, providing useful information about the possibility of interacting with them. (The special case of autonomous agency is considered next.)

This notion of agency relies upon the existence of other agents which provide goals that are adopted in order to instantiate an agent. We must therefore have some agents which can generate their own goals so that there is not an infinite regress in adopting goals. It is these goal-generating agents which are *autonomous*, since they do not depend on the goals of others, but instead possess goals which are *generated* from internal *motivations* which characterise the agent. Motivations are related to goals, but are qualitatively different in that they are not describable states of affairs in the environment. The motivation *greed*, for example, does not specify a state of affairs to be achieved, but may give rise to the generation of a goal to rob a bank. The motivation provides a *reason* for achieving the goal which specifies *what* to do.

In psychology, Kunda [5] informally defines motivation to be, "Any wish, desire, or preference that concerns the outcome of a given reasoning task," and suggests that motivation affects reasoning in a variety of ways including the accessing, constructing and evaluating of beliefs and evidence, and decision making. An autonomous agent is an agent with motivations and some means of evaluating behaviour in terms of the environment and those motivations, so that its behaviour is determined by both external and internal factors. An autonomous agent must be a motivated agent.

Our example robot will be autonomous if it has an internal mechanism for goal generation. For example, with motivations of achievement, hunger and self-preservation, where achievement is defined in terms of riveting panels, hunger in terms of maintaining power levels, and self-preservation in terms of avoiding system breakdowns, the robot will normally generate goals to rivet panels. With low power levels, however, it may instead generate a new goal to recharge its batteries. Alternatively, if it works for too long and is in danger of overheating, it may generate a goal of pausing to avoid damage to its components. In the view described above, the robot is autonomous because its goals are internally generated through motivations in response to its environment.

To present a mathematical description of our definitions, we use the specification language, Z [13], which is increasingly being used in the AI community [4]. However, we are sometimes loose with the syntax when detailing sequences and sets, for example, for the purposes of concision and explication.

An *entity* comprises a set of *motivations*, a set of *goals*, a set of *actions*, and a set of *features* or *attributes* of the entity.

$$
\begin{array}{l}
\underline{\quad Entity\underline{\qquad\qquad\qquad\qquad\qquad\qquad\qquad}} \\
attributes : \mathbf{P}\ Attribute \\
capableof : \mathbf{P}\ Action \\
goals : \mathbf{P}\ Goal \\
motivations : \mathbf{P}\ Motivation \\
\hline
attributes \neq \{\} \\
\end{array}
$$

$$
\begin{array}{l}
\underline{\quad Object\underline{\qquad\qquad\qquad\qquad\qquad\qquad\qquad}} \\
Entity \\
\hline
capableof \neq \{\ \} \\
\end{array}
$$

$$
\begin{array}{l}
\underline{\quad Agent\underline{\qquad\qquad\qquad\qquad\qquad\qquad\qquad}} \\
Object \\
\hline
goals \neq \{\} \\
\end{array}
$$

$$
\begin{array}{l}
\underline{\quad AutonomousAgent\underline{\qquad\qquad\qquad\qquad\qquad}} \\
Agent \\
\hline
motivations \neq \{\} \\
\end{array}
$$

An object, described by its features and capabilities, is just an automaton and cannot engage other entities to perform tasks, nor is it itself used in this way. However, it serves as a useful abstraction mechanism by which it is regarded as distinct from the remainder of the environment, and can subsequently be transformed into an agent, an augmented object, engaged to perform some task or satisfy some goal. Viewing something as an agent means that we regard it as satisfying or pursuing some goal. Furthermore, it means that it must be doing so either for another agent, or for itself, in which case it is autonomous. If it is for itself, it must have generated the goal through its motivations.

For the purposes of the following sections in which we discuss the kinds of relationship that can be modelled, we further refine the agent hierarchy through the introduction of two new terms. A *neutral-object* is an object that is *not* an agent, and a *server-agent* is an agent that is *not* an autonomous agent.

```
┌─ NeutralObject ────────────────────────────────────────────────
│ Object.
├────────────────────────────────────────────────────────────────
│ goals = { } ∧ motivations = { }
```

```
┌─ ServerAgent ──────────────────────────────────────────────────
│ Agent
├────────────────────────────────────────────────────────────────
│ motivations = { }
```

We must consider the multi-agent world as a whole rather than just individual agents. This world is defined below, where all autonomous agents are agents, all agents are objects and all objects are entities. An agent is either a server-agent or an autonomous agent, and an object is either a neutral-object or an agent.

```
┌─ World ────────────────────────────────────────────────────────
│ entities : P Entity
│ objects : P Object
│ agents : P Agent
│ autonomousagents : P AutonomousAgent
│ neutralobjects : P NeutralObject
│ serveragents : P ServerAgent
├────────────────────────────────────────────────────────────────
│ autonomousagents ⊆ agents ⊆ objects ⊆ entities
│ agents = autonomousagents ∪ serveragents
│ objects = neutralobjects ∪ agents
```

3 Engagement and Cooperation

Much existing work has defined cooperation only in terms of helpful agents that are predisposed to adopt the goals of another (e.g.[12, 2]). This assumes that agents are already designed with common or non-conflicting goals that facilitate the possibility of helping each other satisfy additional goals. Our view differs in that autonomous agents will only adopt a goal if it is to their advantage to do so, while non-autonomous agents may benevolently adopt goals. This leads to the distinction between *cooperation* and *engagement* as discussed below.

3.1 Engagement

A direct engagement occurs when a neutral-object or a server-agent adopts some goals. In a direct engagement, an agent with some goals, which we call the *client*, uses another agent, which we call the *server*, to assist them in the achievement of those goals. Remember that a server-agent is non-autonomous, and either exists already as a result of some other engagement, or is instantiated from a neutral-object for the current engagement. No restriction is placed on a client-agent.

We define a *direct engagement* to consist of a client agent, *client*, a server agent, *server*, and the goal that *server* is satisfying for *client*. An agent cannot engage itself, and both agents must have the goal of the engagement.

```
┌─ DirectEngagement ──────────────────────────────────
│ client : Agent
│ server : ServerAgent
│ goal : Goal
├──────────────────────────────────────────────────────
│ client ≠ server
│ goal ∈ (client.goals ∩ server.goals)
└──────────────────────────────────────────────────────
```

The set of all *direct* engagements in the world is given by *dengagement*. For any direct engagement in *dengagement*, there can be no intermediate *direct* engagements of the goal, so there is no other agent, y, where *client* engages y for *goal*, and y engages *server* for *goal*. An agent, c, *directly engages* another server-agent, s, if, and only if, there is a direct engagement between c and s. All of these relationships are given as a set denoted by *dengages*. Finally, the server-agents comprise all agents which are the server agent for some direct engagement and the agents are a superset of those agents which are part of some engagement.

```
┌─ WorldEngagements ──────────────────────────────────
│ World
│ dengagement : ℙ DirectEngagement
│ dengages : Agent ↔ ServerAgent
├──────────────────────────────────────────────────────
│ ∀ eng : dengagement • ¬ (∃ y : Agent; e₁, e₂ : dengagement |
│          e₁.goal = e₂.goal = eng.goal • e₁.server = eng.server ∧
│                  e₂.client = eng.client ∧ e₁.client = e₂.server = y)
│ dengages = {e : dengagement • (e.client, e.server)}
│ serveragents = {d : dengagement • d.server}
│ {d : dengagement • d.server} ∪ {d : dengagement • d.client} ⊆ agents
└──────────────────────────────────────────────────────
```

An *engagement chain* represents a sequence of *direct engagements*. For example, if I use a computer terminal to run a program to access a database in order to locate a library book, then there is a direct engagement of myself and the terminal, of the terminal and the program, and of the program and the database, all with the goal of locating the book. An engagement chain thus represents the goal and all the agents involved in the sequence of direct engagements. In the above example, the agents are: *Me, Terminal, Program, Database*

Specifically, an *engagement chain* comprises some goal, *goal*, the autonomous client-agent that generated the goal, *autoagent*, and a sequence of server-agents, *chain*, where each agent in the sequence directly engages the next. For any engagement chain, there must be at least one server-agent, all the agents involved must share *goal*, and each agent can only be involved once.

```
┌─ EngagementChain ───────────────────────────────────
│ goal : Goal
│ autoagent : AutonomousAgent
│ chain : seq₁ Agent
├──────────────────────────────────────────────────────
│ goal ∈ autoagent.goals
│ goal ∈ ⋃{s : Agent | s ∈ ran chain • s.goals}
│ #(ran chain) = #chain
└──────────────────────────────────────────────────────
```

The set of all engagement chains in the world is given in the schema below

by *engchain*. For every engagement chain, *ec*, there must be a direct engagement between the autonomous agent, *ec.autoagent*, and the first client, of *ec*,
head ec.chain, with respect to the goal of *ec*, *ec.goal*. Further, there must be a
direct engagement between any two agents which follow each other in *ec.chain*
with respect to *ec.goal*. In addition, all the autonomous agents involved in an
engagement chain are a subset of all the autonomous agents.

$$\begin{array}{|l}
\hline
\underline{\quad WorldEngagementChains} \\
WorldEngagements \\
engchain : \mathbb{P}\ EngagementChain \\
\hline
\forall\ ec : engchain;\ s_1, s_2 : Agent \bullet \\
\qquad\qquad (\exists\ d : dengagement \bullet d.goal = ec.goal \wedge d.client = ec.autoagent \\
\qquad\qquad\qquad\qquad \wedge\ d.server = head\ ec.chain) \wedge \\
\qquad\qquad \langle s_1, s_2 \rangle\ in\ ec.chain \Rightarrow (\exists\ d : dengagement \bullet \\
\qquad\qquad\qquad\qquad d.client = s_1 \wedge d.server = s_2 \wedge d.goal = ec.goal) \\
\{ec : engchain \bullet ec.autoagent\} \subseteq autonomousagents \\
\hline
\end{array}$$

Now, in order to additionally specify some other relations between agents, we
will make use of the generic relation *follows*, defined below. This holds between
a pair of elements and a sequence of elements if the first element of the pair
precedes the second element in the sequence.

$$\begin{array}{|l}
\hline
\underline{=\![X]\!=} \\
follows : (X \times X) \longleftrightarrow seq\ X \\
\hline
\forall\ a, b : X;\ s : seq\ X \bullet ((a, b), s) \in follows \Leftrightarrow \\
\qquad\qquad (\exists\ t, u, v : seq\ X \bullet s = t ^\frown \langle a \rangle ^\frown u ^\frown \langle b \rangle ^\frown v) \\
\hline
\end{array}$$

In general, an agent *engages* another agent if there is some engagement chain
in which it precedes the server agent. An agent *owns* another agent, if there is
no other agent using it for a different purpose. In other words, *c* owns *s* if, for
every sequence of server-agents in an engagement chain in which *s* appears, *c*
precedes it or is the autonomous client-agent that initiates the chain. Lastly, an
agent *c directly owns* another agent *s*, if it owns it, and is directly engaging it.

$$\begin{array}{|l}
\hline
\underline{\quad AgentRelations} \\
WorldEngagementChains \\
engages, owns, downs : Agent \longleftrightarrow ServerAgent \\
owns : Agent \longleftrightarrow ServerAgent \\
downs : Agent \longleftrightarrow ServerAgent \\
\hline
engages = \{ec : engchain \bullet (ec.autoagent, head\ ec.chain)\} \cup \\
\qquad\qquad \{ec : engchain;\ c, s : Agent \mid ((c, s), ec.chain) \in follows \bullet (c, s)\} \\
\forall\ c : Agent;\ s : Agent \bullet (c, s) \in owns \Leftrightarrow (\forall\ ec : engchain \mid s \in ran\ ec.chain \bullet \\
\qquad\qquad ec.autoagent = c \vee ((c, s), ec.chain) \in follows) \\
\forall\ c : Agent;\ s : Agent \bullet (c, s) \in downs \Leftrightarrow (c, s) \in owns \cap dengages \\
\hline
\end{array}$$

3.2 Cooperation

Two autonomous agents are said to be *cooperating* with respect to some goal
if one of the agents has adopted goals of the other. This notion of autonomous

goal acquisition applies both to the *origination* of goals by an autonomous agent for its own purposes, and the *adoption* of goals from others, since in each case the goal must have a positive motivational effect [7]. For autonomous agents, the goal of another can only be adopted if it has such an effect, and this is also exactly why and how goals are originated. Thus goal adoption and origination are related forms of goal generation.

That is to say that the term *cooperation* can be used only when those involved are autonomous and, at least potentially, capable of resisting. If they are not autonomous, nor capable of resisting, then one simply *engages* the other. The difference between engagement and cooperation is in the autonomy or non-autonomy of the entities involved. It is senseless, for example, to consider a terminal cooperating with its user, but meaningful to consider the user engaging the terminal. Similarly, while it is not inconceivable for a user to engage a secretary, it makes better sense to say that the user and the secretary are cooperating, since the secretary can withdraw assistance at any point. This applies at the level of the definitions in the model, and in real situations. Cooperation is thus a *symmetric* relation between two autonomous agents. Engagement, by contrast, is asymmetric between a server-agent and another client agent.

A *cooperation* describes a goal, the autonomous agent that generated the goal, and those autonomous agents who have adopted that goal from the generating agent. Thus in this view, cooperation cannot occur unwittingly between agents, but must arise as a result of the motivations of both of the individuals involved.

```
┌─ Cooperation ─────────────────────────────
│ goal : Goal
│ generatingagent : AutonomousAgent
│ cooperatingagents : P AutonomousAgent
├───────────────────────────────────────────
│ #cooperatingagents ≥ 1
│ ∀ aa : cooperatingagents • goal ∈ aa.goals
│ goal ∈ generatingagent.goals
└───────────────────────────────────────────
```

The set of all cooperations in the world is given by *cooperations*. Further, we say that agent x_1 *cooperates* with agent x_2 if and only if x_1 and x_2 are autonomous, there is some cooperation in which either x_1 or x_2 is the agent that generated the goal, and x_2 or x_1 respectively is one of the cooperating agents. The set of all such relationships is given below by *cooperates*. We assert also that the relationship is symmetric — it is equal to its own inverse — since if agent x_1 is cooperating with x_2 then, necessarily, x_2 is cooperating with x_1. Thus order in cooperation is not relevant.

```
┌─ WorldCooperations ───────────────────────
│ World
│ cooperations : P Cooperation
│ cooperates : AutonomousAgent ↔ AutonomousAgent
├───────────────────────────────────────────
│ cooperates = ⋃{a1, a2 : AutonomousAgent |
│       (∃ c : cooperations • a1 = c.generatingagent ∧
│           a2 ∈ c.cooperatingagents) • {(a1, a2), (a2, a1)}}
│ cooperates~ = cooperates
└───────────────────────────────────────────
```

4 Agent Models

The framework described above provides useful structure that can be exploited by intelligent agents for more effective operation. However, this is only possible if each agent maintains a model of their view of the world. Specifically, each agent must maintain information about the different entities in the environment, so that both existing and potential relationships between those entities may be understood and consequently manipulated as appropriate.

For example, objects are not involved in any relationship with agents. If they were serving some useful purpose at a particular point in time, then they would be viewed as agents. The view of an entity as an agent thus indicates that it is *engaged* by another agent, either directly or back through a chain of intermediate agents, grounded with an autonomous agent at the head of the chain. Knowledge of the agency of an entity allows us to reason about its function and the agent or agents that are engaging it. Thus, I understand that my colleague's pencil is her agent for writing, and that if I want to use it I must negotiate with her as the engaging agent. Alternatively, if one robot is assisting another to move a crate of books, my models of the two robots may provide information as to the direction of the relationship. That is to say that if I have a model of one as an agent and another as an autonomous agent then, subject to circumstances, it may be sensible for me to infer that the non-autonomous agent is engaged by the autonomous agent. In this case, I must negotiate with the autonomous robot if I want either of the two agents to help me in my efforts to achieve my own goals.

In this example, this agency information provides useful structure which allows us to consider issues such as whether it is possible to negotiate with another, if such negotiation would be likely to achieve the desired result, how the negotiation might be approached, and so on.

Returning to the example of the pencil above, we can see how the richness of the situation can be captured through agent models. If my colleague is using her pencil, then I can view it as an agent satisfying the goal of writing some notes, for example, for its engaging agent, my colleague. Now, if I want to use her pencil, I can reason about the situation by considering the goal that the pencil is satisfying. First, I know that the goal was generated by my colleague, and I must therefore negotiate with her to secure use of the pencil. However, my knowledge of her motivations that generated the goal may lead me to decide that I will not succeed in securing use of the pencil for any number of reasons. If I know that the goal was generated because of an imminent important deadline, then I may decide that I will not be successful. Alternatively, if I know that the motivation for using the pencil was weak, then I may rate my chances of success highly.

It is important to note that the word, "pencil", of this example might just as easily be replaced by the word, "robot". The relationships described here are exactly the same.

4.1 Motivation

The key to the previously defined relations is motivation. Motivation is the 'force' that causes engagement chains to built up, satisfying goals that mitigate the motivation. In attempting to understand the nature of the relationships between entities in the world, and using or augmenting them, it is therefore necessary to be able to assess the relative strengths of motivation that caused those engagements. Similarly for cooperation, though here motivation plays an even greater role, since cooperation is symmetric and requires an assessment of motivation in all cooperating entities. In this paper, we are concerned not with investigating motivation, but in using it to describe engagement and cooperation. Related work has explored the nature of motivation and similar concerns in more detail [11, 9, 8], and we will only provide a simplified model here.

In order to retrieve goals to mitigate motivations, an autonomous agent must have some way of assessing the effects of competing or alternative goals. Clearly, the goals which make the greatest positive contribution to the motivations of the agent should be selected. The *AssessGoals* schema below describes how an autonomous agent monitors its motivations for goal generation. First, the *AutonomousAgent* schema is included, and a new variable representing the repository of available known goals, *goalbase* is declared. Then, the motivational effect on an autonomous agent of satisfying a set of new goals is given. The *motiveffect* function returns a numeric value representing the motivational effect of satisfying a set of goals with a particular configuration of motivations and a set of existing goals. The predicate part specifies all goals currently being pursued must be known goals that already exist in the goalbase. Finally, for ease of expression, we define a function related to *motiveffect* called *satisfy* which returns the motivational effect of an agent satisfying some goals.

$$
\begin{array}{l}
\underline{\quad AssessGoals \underline{\hspace{8cm}}} \\
AutonomousAgent \\
goalbase : \mathbf{P}\ Goal \\
motiveffect : \mathbf{P}\ Motivation \longrightarrow \mathbf{P}\ Goal \longrightarrow \mathbf{P}\ Goal \longrightarrow \mathbb{Z} \\
satisfy : \mathbf{P}\ Goal \longrightarrow \mathbb{Z} \\
\hline
goals \subseteq goalbase \\
\forall gs : \mathbf{P}\ goalbase \bullet satisfy\ gs = motiveffect\ motivations\ goals\ gs
\end{array}
$$

Therefore, $satisfy_x(gs)$ is the motivational effect on the autonomous agent x of satisfying the goals, gs. We further define the *complement* of a goal, which we write $\overline{goal}$, to be the goal to prevent that *goal* being achieved.

Consider the following situation in which, according to the model of some autonomous agent y, the autonomous agent x is directly engaging the server agent z for some goal g_x so that:

$(x, z, g_x) \in dengagement$

and further suppose that agent y wants to use z for some other goal g_y. There are several possible courses of action for y.

- y can persuade x to share z
- y can persuade x to release z.

- y can attempt to take z by force without x's permission.
- y can give x priority and find an alternative.

Any decision as to which alternative to take requires an analysis of both y's motivations and x's motivations. The analysis of their motivations below is solely from y's point of view.

- $satisfy_x\{g_y, g_x\} > satisfy_x\{g_x\}$. If g_y and g_x do not conflict, it is possible for z to adopt both of the goals of y and x without violating any motivational constraints. So long as the motivational effect on x of satisfying both goals is more than satisfying just her own, x will be disposed to share z.
- $satisfy_x\{g_y\} > satisfy_x\{g_x\}$. y understands that x stands to gain more from enabling y to satisfy y's goal than from satisfying its own goal. This is due to the effect that a positive change in y's motivations will have on x. This may require that y explains and persuades x of the degree of effect that g_y will have. For example, a friend may currently be reading a book that I want to borrow. My goal of borrowing the book may conflict with my friends goal, but because she wants to please me and does not need to read the book now, she happily lends the book to me.
- $satisfy_y\{g_y\} > satisfy_y\{\overline{g_x}\}$. It may seem obvious that the motivational effect on y of satisfying its goal should be greater than the motivational effect on y of satisfying x's goal. However, if x's goal is not satisfied, then the motivational effect on x will be negative, and this results in a state which must be considered in terms of its effect on y. (In other words, a negative motivational effect on y, particularly if it was a consequence of x, may result in a negative motivational effect on x.) Thus this alternative may be chosen if there is a positive motivational effect from y's goal being satisfied, and this is greater than the negative consequences of x's goal not being satisfied. Note that this relies on the relationship of y to x. Normally, y's motivations will be such that negative motivational effect on other agents will lead to some negative motivational effect on y itself. If, however, y is motivated by malicious concerns, then it is certainly possible that the consequences of x's goal not being satisfied may have a positive motivational effect on y. While we do not envisage such a situation arising regularly, and though this is a case typically not considered in related work, it ought to be possible within any formalism. By using motivations in the way we describe, we allow the possibility of perverse configurations leading to such malicious behaviour, but envisage appropriate design of motivations so that this does not arise. In summary, this captures normal social behaviour by which we act so as to avoid annoying others, but allows for situations where we may deliberately choose to annoy them. For example, if my friend is reading a book that I want to borrow, then I can simply take the book from her without permission. It only makes sense for me to do this, however, if the benefit I get from having the book is more significant than the bad feeling caused in my friend by my having taken it forcibly.

– $satisfy_y\{g_y\} < satisfy_y\{\overline{g_x}\}$. y understands that the motivational effect of satisfying its goal will be less than the effect of causing a negative motivational effect on x through g_x nor being satisfied. This affects y's behaviour, because its motivations are configured in such a way that y is concerned for x. Like the previous case, this describes the situation where we do not act if that action is likely to annoy others.

4.2 Applying Agent Models

Consider the situation in a library in which a user wants to locate a particular book. The autonomous agents in the library world include the user and the librarian. Now, suppose that in order to locate the book, the user requires the assistance of the librarian. Since the librarian is autonomous, this means that the user must persuade the librarian to *cooperate* with her in achieving her desired goal. In attempting to locate the book for the user, the librarian uses a terminal to invoke a computer program which, in turn, accesses the library databases and performs the relevant query. These relationships can be captured very easily in terms of the predicates defined earlier.

First, the librarian, L, cooperates with the user, U.

$(L, U) \in cooperates$

Then, the librarian uses the terminal, T, the program, P, and the database, D, to locate the book.

$(L, T) \in downs \wedge$
$(T, P) \in dengages \wedge$
$(P, D) \in dengages \wedge$
$\Rightarrow (L, D) \in engages$

In this scenario, the librarian owns (and also directly engages) the terminal, T, and this engagement makes T a server-agent. In turn, T directly engages P, making it a server-agent, and similarly, P engages D. This engagement chain is constructed through the engagement of agents by other agents in order to satisfy the goal of locating the book, and is shown in Figure 2.

Let us now suppose that the terminal can only be used by one person at a time, but the program and database can be used by many people at once. With this information, other agents in the library can effectively decide what courses of action to follow to achieve their aims. Another user, for example, $U2$, may also want to locate a book. $U2$ must decide what action to take, or agents to invoke, on the basis of the motivations of the autonomous agents involved in the engagement. Since the terminal cannot be used by multiple agents and is *owned* by the librarian, $U2$ cannot share it, but can choose to take it forcibly, to persuade L to release it, or to give L priority and either wait or find an alternative terminal. Note that this last possibility is not constrained by the program and database being engaged by L, since they are not owned by L. If $U2$ attempts to take the terminal forcibly, the consequences may be severe in terms of detrimental motivational effect on L. Persuading L to release the terminal requires another cooperation, between L and $U2$, which may not be

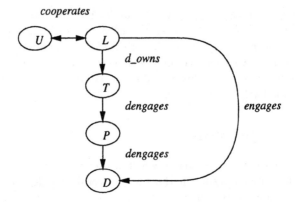

Fig. 2. An Engagement Chain for the Library Scenario

forthcoming if the motivational effect of $U2$'s goal on L is less than L's current goal. One of the alternative options is for $U2$ to ask another librarian, $L2$, to locate the book, while another solution would be for $L2$ to take part in separate cooperations with $U2$ and L so that the terminal is released for use in finding the second book. This is because the negative motivational effect on L of not cooperating with $L2$ will be significantly greater than the negative motivational effect of not cooperating with the user $U2$.

Figure 3 shows another situation that arises when $U2$ asks another librarian, $L2$, to locate a book, and $L2$ uses a different terminal, $T2$. Note that in order to avoid complicating the diagram, not all of the relations between the entities are shown, though they can be inferred. For example, L engages T and P and D, while $L2$ engages $T2$ and P and D.

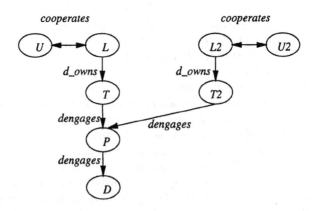

Fig. 3. Additional Engagement and Cooperation in the Library Scenario

5 Related Work

This research shares some common themes with the concepts of dependence networks [10]. Dependence networks use *external descriptions* which store information about other agents, comprising a set of goals, actions, resources and plans. The goals are those an agent wants to achieve, the actions are those an agent is able to perform, the resources are those over which an agent has control, and the plans are those available to the agent, but using actions and resources which are not necessarily owned by the agent. This means that one agent may *depend* on another in terms of actions or resources in order to execute a plan.

Using external descriptions, the authors distinguish three different forms of autonomy, as follows. An agent is *a-autonomous* if for a given goal, according to a set of plans, there is some plan in the set which consists of only actions of that agent. An agent is *r-autonomous* if, similarly, any resources needed belong to the agent. An agent is *s-autonomous* if it is autonomous in both of the above ways (and does not, therefore, need any help). If agents are not autonomous, then they may *depend*, for resources or actions, on other agents. By contrast, our work describes autonomy not as action or resource dependence, but as the ability to make one's own choices, to generate goals. This is a far stronger notion of autonomy. The fact that a pocket calculator has the resources and the actions necessary for adding some numbers surely does not make it autonomous. A more complete analysis of this work, and a reformulation of it in our framework can be found in [3].

Moreover, even though this kind of categorisation is useful, it is not clear how much help it would be to a motivated agent. An agent may believe that another agent will perform a required action for reasons of continuing friendship, trust, promise, and so on. The motivational context of these dependencies is what is important, and the target agent may not care to investigate whether there are any specific actions that it has promised to perform for an agent.

While our work has concentrated on defining the relationships that are involved in cooperation, for example, other related work has addressed the *processes* of cooperative problem solving, also in a formal way [15]. One possible way forward might be to investigate the possibility of integrating these two approaches.

6 Conclusions and Further Work

The work reported in this paper is part of a broader project with the aim of developing effective mechanisms for autonomous communication. To date, we have developed a formal framework structuring the world into object, agent and autonomous agent components. Based on those definitions, we are developing mechanisms for goal generation and transfer or adoption so that agents may interact with each other in useful ways. In particular, we describe and specify the differences that arise in these interactions which we characterise as *engagements* of non-autonomous agents, and *cooperation* between autonomous agents. The next objective is to incorporate explicit mechanisms for deliberation, which will include both planning and communication. Our immediate aim is to provide a communication facility through the use of *knowledge interchange protocols* [1]. Finally, we intend to add a learning component which will dynamically acquire knowledge of the environment and of protocols so that more efficient behaviour can be obtained.

References

1. J. A. Campbell and M. P. d'Inverno. Knowledge interchange protocols. In Y. Demazeau and J. P. Muller, editors, *Decentralized Artificial Intelligence*. Elsevier North Holland, 1989.
2. P. R. Cohen and C. R. Perrault. Elements of a plan-based theory of speech acts. *Cognitive Science*, 3(3):177–212, 1979.
3. M. d'Inverno and M. Luck. A formal view of social dependence networks. In *Proceedings of the First Australian DAI Workshop*. Springer Verlag, To appear, 1996.
4. R. Goodwin. Formalizing properties of agents. Technical Report CMU-CS-93-159, Carnegie-Mellon University, 1993.
5. Z. Kunda. The case for motivated reasoning. *Psychological Bulletin*, 108(3):480–498, 1990.
6. M. Luck and M. d'Inverno. A formal framework for agency and autonomy. In *Proceedings of the First International Conference on Multi-Agent Systems*, pages 254–260. AAAI Press / MIT Press, 1995.
7. M. Luck and M. d'Inverno. Goal generation and adoption in hierarchical agent models. In *AI95: Proceedings of the Eighth Australian Joint Conference on Artificial Intelligence*. World Scientific, 1995.
8. D. Moffat and N. H. Frijda. An agent architecture: Will. In *Proceedings of the 1994 Workshop on Agent Theories, Architectures, and Languages*, 1994.
9. T. J. Norman and D. Long. A proposal for goal creation in motivated agents. In *Proceedings of the 1994 Workshop on Agent Theories, Architectures, and Languages*, 1994.
10. J. S. Sichman, Y. Demazeau, R. Conte, and C. Castelfranchi. A social reasoning mechanism based on dependence networks. In *ECAI 94. 11th European Conference on Artificial Intelligence*, pages 188–192. John Wiley and Sons, 1994.
11. A. Sloman. Motives, mechanisms, and emotions. *Cognition and Emotion*, 1(3):217–233, 1987.
12. R. G. Smith and R. Davis. Frameworks for cooperation in distributed problem solving. *IEEE Transactions on Systems, Man and Cybernetics*, 11(1):61–70, 1981.
13. J. M. Spivey. *The Z Notation*. Prentice Hall, Hemel Hempstead, 2nd edition, 1992.
14. M. J. Wooldridge and N. R. Jennings. Agent theories, architectures, and languages: A survey. In *Proceedings of the 1994 Workshop on Agent Theories, Architectures, and Languages*, 1994.
15. M. J. Wooldridge and N. R. Jennings. Formalizing the cooperative problem solving process. In *Proceedings of the Thirteenth International Workshop on Distributed Artificial Intelligence*, 1994.

A Methodology for Developing Agent Based Systems

Elizabeth A. Kendall, Margaret T. Malkoun, and Chong H. Jiang
kendall@rmit.edu.au, maggie@rmit.edu.au, and jiang@rmit.edu.au
Department of Computer Systems Engineering
Royal Melbourne Institute of Technology

Abstract

Before agents can be used extensively a methodology must be established for the development of agent based systems. This methodology must encompass modeling or analysis, design, and implementation of both the technical and organisational aspects of the systems. This paper outlines a methodology for the engineering of agent based systems. The methodology is based upon the IDEF (ICAM Definition) approach for workflow modeling and the use case driven approach to object oriented software engineering. The methodology is illustrated via a case study in the area of discrete parts manufacturing.

1 Introduction

As with any area of information technology and engineering, an agent based methodology must encompass the full product life-cycle: modeling or analysis, design, implementation, and maintenance.

Some initial work in this area has been completed, applying and extending conventional models to agents. Work by (Huntbach, Jennings, and Ringwood, 1995) indicated that an agent could be identified as a process or a data store in a data flow diagram with real time extensions. However, agent behaviour is actually quite different from a data flow process or a data store. The need to consider *work flow and not just data flow* (Klein, M., 1995) for agents has been discussed, as agents cooperate and coordinate their activities much as individuals do in organisations or enterprises. (Klein, M., 1995) provides the following table for a comparison of dataflow and workflow representations and states that dataflow representations do not meet the needs of agent based systems; workflow representations are found to be more suitable.

Dataflow	Workflow
Focus on communication between computer systems	Focus on business processes with human participants
Pass information	Pass tasks
Pre-defined and rigid	Can have exceptions
First in First Out queues	Task queues and reasoning about ordering/ merging
Systems are interchangeable	Agents have unique skills and positions
Requires input data	Requires task execution environment

In this paper, existing methodologies for workflow and object oriented modeling and analysis are extended to encompass agent oriented systems. An object oriented methodology is employed, but many aspects of the object oriented model have an equivalent workflow representation that is discussed. In particular, the IDEF methodology for integrated computer aided manufacturing (ICAM) and the use case driven object oriented software engineering (OOSE) approach are utilised. These tools have been employed because they are widely used and best capture the work flow and active nature of an agent, surpassing approaches that only address data flow or work flow.

2 Background
2.1 Agents
2.1.1 Definitions

There are two views of agents: weak and strong (Wooldridge and Jennings, 1995). A weak definition of agency is: i) autonomous - an agent has goals and plans for achieving it goals. Agents operate without direct human or other intervention, ii) social - an agent can interact with other agents, iii) reactive - an agent can perceive its environment and respond to changes, and iv) pro- active - an agent can affect its environment. A strong definition of agency also has one or more of: v) mentalistic notions - an agent has beliefs, desires, intentions, obligations, commitments, and choices, vi) rationality - an agent performs actions which further its goals, vii) veracity viii) adaptability.

Strong agents are emphasised in this research, although only items i) through vi) above are considered. Strong agents reason about their knowledge and beliefs to select a plan that could achieve their stated goals. Beliefs are primarily represented in extensions to first order predicate calculus (Georgeff and Ingrand,1990)(Brenton, 1995) and KIF (Knowledge Interchange Format) (Genesereth, Fikes, *et al.*, 1992).

Agent pro-active behaviour is represented in terms of a plan library. An agent selects a plan from the plan library on the basis of the goals that are to be accomplished. A plan is instantiated when a triggering event occurs that satisfies its invocation condition, such as the posting of a new goal. In addition to the invocation condition, context conditions must also be considered before instantiation. An instantiated plan is an intention. The body of a plan is a set of tasks that can be sub-goals and actions. When an intention is formed, these tasks are executed by the agent in an effort to reach the stated goal.

2.1.2 Agent Cooperation

Agents are meant to operate in a distributed environment, cooperating with other agents. Therefore, agents also require behaviour that allows them to communicate with other agents. KQML (Knowledge Query and Manipulation Language) (Finin, Weber, *et al*, 1993) is a language for communication among agent based programs. With KQML, agents can communicate in a structured format.

Agents must do more than communicate; they must be able to cooperate and negotiate with each other. When agents negotiate, they engage in protocols that can be represented in speech acts and state machines, and languages have been developed to support this. COOL is an extension to KQML (Barbuceanu and Fox, 1995), and with it agents can make proposals and counterproposals, accept and reject goals, and notify other agents of goal cancellation, satisfaction, or failure. Another language, AgenTalk (Kuwabara, Ishida, and Osato, 1995), has a specialisation mechanism for introducing new protocols as extensions of predefined ones. Whereas COOL is a general purpose language for negotiation, AgenTalk allows specialised protocols to be developed on the basis of existing ones by adding states and transitions.

2.2 Object Oriented Software Engineering

In the simplest view, an object is a representation that has properties and behaviour. In a more complete view, the notion of an object encompasses all of the following features (Booch, 1994): compound objects, inheritance, encapsulation, identity, state and behaviour.

During object oriented analysis and design, an object oriented model is developed for a given application. The Object Modeling Technique (OMT) (Rumbaugh *et al.*, 1991), consists of three complementary models: the object model, the dynamic model, and the functional model. The object model shows every object in the application and static relationships between objects. The dynamic model includes scenarios, event traces where objects interact with one another, and state transition diagrams. The functional model resembles a data flow diagram.

Dynamic and functional modeling are only carried out to enhance the object model. In dynamic modeling, state transition diagrams are used to provide an overall view of an object's dynamic behaviour and to gain insight into inheritance. Subclasses inherit the state diagram of a superclass, and a subclass can only add states and transitions. Therefore, the protocols of AgenTalk (section 2.1.2) are in fact classes of objects that can be specialised to new classes using inheritance, as stated in (Kuwabara et al., 1995).

The use case approach, also called OOSE (Object Oriented Software Engineering) (Jacobson, 1992) offers capabilities that are relevant to partitioning. A use case is a description of how users interact with the system in a certain mode of operation. A user in a particular role is known as an actor, and each actor *uses* one or more use cases. Objects relevant to a given use case are then identified, and their dynamic behaviour within the use case is analysed with event traces. Use cases can be abstracted, specialised and extended conditionally and a scenario is an instance of a use case.

2.3 Workflow and Enterprise Modeling Methodologies

Methodologies have been proposed for workflow and enterprise modeling. The IDEF (ICAM Definition) workflow method (Bravoco and Yadav, 1985a and 1985b) has been widely adopted as a standard. The IDEF method comprises three modeling projections or views: the function model (IDEF$_0$), information model (IDEF$_1$), and dynamic model (IDEF$_2$). In the function model, the organisation is described in terms of a hierarchy of functions that are decisions, activities, or actions. Each function is given an ICOM (Input, Control, Output, and Mechanism) representation. The IDEF$_0$ model corresponds to a workflow representation.

The IDEF$_0$ models can be taken to a high degree of detail through deeper levels of diagrams. An example is shown in Fig. 1. Here, an IDEF$_0$ function includes decisions or choices that are made on the basis of control input. The Decide Action function chooses a method or activity (Method 1 or Method 2 in the diagram), based upon the Input and the Control Input, which may in fact be meta-level or strategic knowledge.

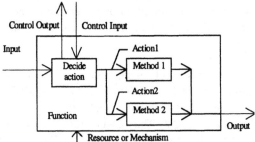

Fig. 1. Detailed Function Model with Choices in IDEF$_0$

3 Approach

This section provides an overview of how object oriented models can be translated into IDEF enterprise models and vice versa. It also discusses the correspondences between objects and agents. The net result is a description of a methodology for how agents can be identified from IDEF or use case models and then further analysed via object oriented and IDEF techniques. Both IDEF and object oriented techniques are addressed to capture the workflow and active nature of agents.

3.1 Object Oriented and Enterprise Modeling Methodologies

The three levels or views of the IDEF methodology are analogous to the three projections employed in object oriented software engineering. If the IDEF$_0$ function model is re-expressed in terms of an OMT functional model, active and passive objects can be identified. Once these objects have been identified, they are then refined in the information and dynamic models.

When use case diagrams and IDEF$_0$ diagrams are completed to the same level of detail, IDEF functions appear directly as a use case. A use case groups a behaviourally related sequence of transactions. The corresponding IDEF$_0$ diagram would show the same transactions, and it would include input, output, and control flow. In an IDEF$_0$ diagram, use case actors would appear as resources or mechanisms. The objects' interactions are depicted in the use case interaction diagrams.

When actors and control objects manage a given use case, they are generating or outputting control information and passing it on as input to other components of the system. In this way, policy or decision making actors carry out IDEF functions with control output.

3.2 Agents and Objects
3.2.1 The Difference Between Agents and Objects
Agents are real time, autonomous objects that carry out pro-active behaviour. An object (concurrent objects are active objects) sequentially executes methods or member functions, whereas a strong agent encompasses automated reasoning that allows it to choose one alternative. The distinctions between objects and agents are the following:

Agents and objects:
- *Compound* - can have properties or attributes that are objects or agents themselves.
- *Encapsulation* - can hide or protect data and behaviour
- *Identity* - have identity that is not determined solely by their attribute values.
- *State* - behaviour is influenced by their history.
- *Behaviour* - can act and react through receiving and sending messages
Objects:
- *Inheritance* - types can have subtypes but synchronisation behaviour can not be inherited.
- *Unstructured messages* - can communicate through unstructured messages
Agents:
- *Reasoning* - can perform automating reasoning to "intelligently" choose between options
- *Pro-active behaviour* - have beliefs, knowledge, goals, plans, and intentions
- *Concurrency* - can act in a distributed, cooperative manner that must be accomplished in a synchronised, deterministic fashion in real time
- *Structured messages* - can communicate with each other through structured messages

---

3.2.2 Objects as Beliefs
The first level of the relationship between agents and objects is beliefs. Passive objects that store attributes frequently appear in belief databases (Georgeff and Ingrand, 1990). The Ontolingua project (Gruber, 1992) has produced a software system that translates frame based knowledge representations into a KIF beliefs database, and vice versa. Passive domain objects that appear in an agent's belief database are similar to frames. Ontolingua unifies frame formalisms and predicate calculus representations within a coherent conceptualisation. With this capability, Ontolingua can directly translate the static properties of objects that appear in an agent's belief database into KIF and other predicate calculus representations.

3.2.3 Objects as Sensors and Effectors
Objects with encapsulation and dynamic behaviour can be utilised by agents to carry out pro-active behaviour. In this regard, objects can be sensors and effectors for agents (Rao, 1995). One object will serve as a sensor or perceptor for the agent, while another is employed as an effector or actuator.

The sensor object monitors and filters output from objects in the application, alerting the agent when certain conditions exist. Effector objects can carry out an agent's intentions and actions, impacting other objects in the environment. The effector object takes the agent's actions and relays them to other objects within the application.

The sensor- agent- effector team works together, in a pattern (Booch, 1994)(Gamma *et al.*, 1995). The sensor watches the external environment, while the agent makes decisions, and the effector brings about the changes desired by the agent.

3.2.4 Actors, Agents, and Use Cases
Use cases depict how humans and/ or organisations enter and interact with each other and with a system. One use case represents one mode of operation for the system, with each actor as a user(s) in a certain role. Agents enter applications to augment or automate human reasoning and collaboration. As such, agents replace or directly support actors in one or more given use case.

For a full requirements analysis, use cases must be specified for each scenario of the application, and a use case maps to an $IDEF_0$ function, as discussed above. Additionally, at this level of detail each use case extension maps to an actor's (or agent's) plan. Conditional extends relationships --- when one use case is inserted into another one if certain circumstances exist --- correspond to context or invocation conditions that determine when an agent's plan is to be intended. Some control object behaviour also maps to plans, whereas other behaviour will map to sensors and effectors. The concept that use case extensions may be additive is analogous to more than one plan being intended at one time (Carroll, 1995).

When two or more actors appear in a use case, they collaborate or cooperate via their own message passing and negotiation. In an IDEF representation, resources or individuals collaborate either by appearing in the same functional block or by exchanging input and output, with one function's output serving as control input to another function, and perhaps vice versa. In the agent based system, the actors and resources become the agents and their negotiation and information exchange becomes a coordination protocol (section 2.1.2).
Use cases with one or more actors can be abstracted and specialised. At the correct level of abstraction, the type of behaviour exhibited by the actor(s) may be recognised as belonging to that of a certain type of single function agent (Dunskus et al., 1995). Additionally, some of the interchange between two of the actors in the event trace can be identified as an abstract or specialised coordination protocol.

3.3 A Proposed Agent Oriented Methodology
Based on the discussions above, agents appear in applications when pro-active behaviour --- automated decision making --- is required. Pro-active behaviour is necessary when control decisions are required to augment or replace human decision

making. Agents therefore arise in use cases to replace or assist actors. System components or use cases that do not require automated decision making or pro-active behaviour do not feature any agents.

Applications include modules or functions where decision making is carried out. In particular IDEF functions with control information as output capture the decision making components of an enterprise.

The key results are summarised in Fig. 2 where Fig. 1 (an IDEF function model with control output) is modified to indicate the corresponding agent representation. The inputs and outputs from the function are the same, but the interior to the functional block now shows an agent (rounded box). The agent has beliefs, goals, plans, and an interpreter, with passive objects shown as beliefs. Instantiated plans become intentions, through the interpreter's processing.

Three kinds of intentions are indicated as an example for illustration. Task 1 represents coordination/ collaboration with other agents, and task 3 represents tasks to be carried out by the effector. Tasks 2 and 4 are a goal and a fact for the Interpreter. A sensor and effector are also shown, with the sensor taking input from an object and the effector impacting an object.

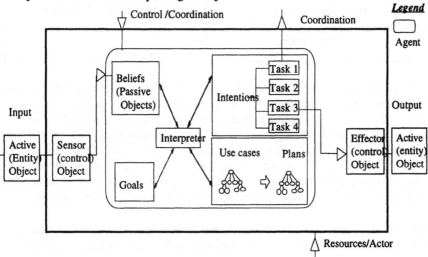

Fig. 2. The Correspondence between Enterprise Activities, Use Cases and Agents

Comparing Fig. 1 with Fig. 2, the 'Decide Action' has been replaced with the 'Interpreter' that decides on an action based on invocation and context condition. The methods or actions are in the tasks which are intended plans. The module in Fig. 2 can negotiate with other modules and then decide on actions to carry out the tasks; thus it is an intelligent module. This negotiation is shown as a control output.

The correspondence between IDEF and use case model components and agent system components is illustrated in the following table and in the discussion that follows it.

IDEF Model	Use Case (OOSE)	Agent Oriented System
Function with Control Output	Use Case and Use Case Extension	Goal and Plan
Resource	Actor	Agent
Functional Input	- Entity object - Actor	- Coexisting object via sensor - Belief
Control Input	- Entity object - Actor input - Conditional use case extension	- Coexisting object via sensor - Agent - Context or invocation condition - Meta-level knowledge
Control Output	- Control object	- Goal or sub-goal - Effector to coexisting object
Functional Output	- Control object	- Effector to coexisting object
More than one resource per function or information exchange between resources	- More than one actor per use case - Use case event trace - Use case abstraction and specialisation	- Agent collaboration - Coordination protocol - Coordination protocol abstraction and specialisation - Single function agent behaviour
Information Model	Entity objects	- Belief - Coexisting object

Only $IDEF_0$ functions with control output are of relevance to agent oriented systems; they correspond to a use case and use case extensions and to an agent's goals and plans. The agent itself appears as a resource or mechanism in the IDEF model and as an actor in the OOSE approach. In an agent oriented system, functional input appears as beliefs or as information from a coexisting object via a sensor. The source of control input is an actor or an entity object, and it determines which use case extension, if any, is followed. In an agent oriented system control input comes from a coexisting object via a sensor, another agent, or meta-level knowledge.

Control output in an IDEF sense corresponds to control object output in OOSE, and this output can be targeted for entity objects or to the actor. An agent's control output can be represented as a goal or sub-goal, and it is transmitted to a coexisting object via an effector.

When more than one actor appears in a use case, or when two or more IDEF resources are involved in a function or exchange information to carry out functions, the agents are engaged in cooperation. The coordination follows a protocol, and the sequence or script would also appear in the event trace for the use case. Use cases and scripts can be abstracted, leading to the identification of single function agent behaviour and known protocols. Lastly, the IDEF information model contains the entity objects of OOSE and the beliefs and coexisting objects of the agent oriented system.

Once types of objects and the agent itself have been identified, object oriented software engineering can be carried out to further define the objects. Further analysis and design of the agent system must also be accomplished. Beliefs and knowledge can be generated directly from any passive entity objects via Ontolingua or similar translation software.

4 Case Study

4.1 Overview

The methodology is illustrated in a case study from discrete parts manufacturing. Customers place orders for parts of given shapes, sizes, and qualities, and the parts are selected for processing and then grouped into batches. There are high quality batches and normal quality batches. There are two kinds of machines: normal and high quality. Material to be used for the parts is of varying dimensions, and material availability is a consideration. The time to process an order is based upon material, and part handling and forming time. Forming time is dependent upon part dimensions and quality, and any complicated features of the part.

4.2 IDEF Model

$IDEF_0$ function model diagrams are provided in Figs. 3, 4, and, 5 at levels 1, 2, and 3, respectively. Without cutting, the system breaks down into five basic functions: Create Parts, Select Parts and Sheets, Create Batch, Assign Machine, and Determine and Evaluate Cost.

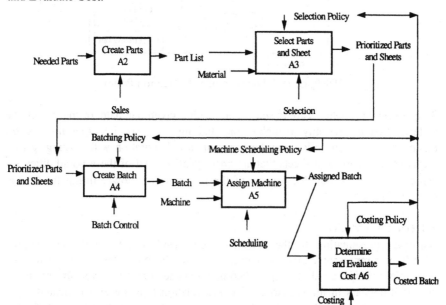

Fig. 3. $IDEF_0$ Function Model Overview of Case Study Level 1 Diagram.

Determination of the part selection policy, batching policy, machine scheduling policy, and the costing policy are addressed in level 2 $IDEF_0$ diagrams, such as the one provided in Fig. 4. In Fig. 4, part selection policy is determined on the basis of parts and material. Information supplied from the control output of costing is also used, along with strategic knowledge. The selection policy is then used to control part and sheet selection; the selection policy decision may also impact costing control input. Determining batching policy and machine scheduling policy would be treated in the same way.

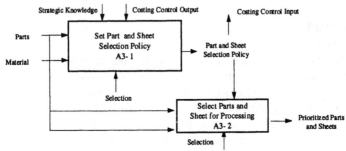

Fig. 4. Select Parts and Sheet $IDEF_0$ Level 2 Diagram

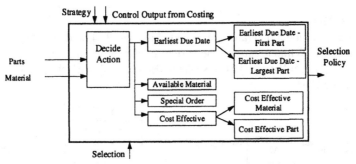

Fig. 5. $IDEF_0$ Level 3 Diagram of Set Selection Policy

Simple selection policies are to select parts with a given due date, or that match the material that is presently available for handling, or a special order may occur. Lastly, the selection policy may be based upon collaboration with costing, giving precedence to parts and sheets that are *cost effective*. All of these possibilities are depicted in Fig. 5; detailed aspects of the function are shown, fitting the format depicted in Fig. 1.

4.3 Use Case Model
Use case representations for the case study are shown in Fig. 6. In the diagram, the stick figures represent actors, while each labelled ellipse or oval depicts a use case. *Uses* relationships are shown by dashed double headed arrows; an actor uses a use case if it appears in the scenarios. *Extends* relationships are also indicated (by a solid and single headed arrow), and one use case extends another if it adds further

steps to the scenario under certain conditions. If an actor *uses* a base use case it also uses the use case extensions, so the double arrows are not repeated.

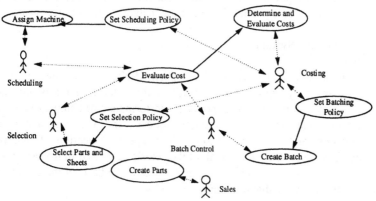

Fig. 6. Use Cases for the Application

The top level use case diagram of Fig. 6 depicts the nine functions (five level 1 and four level 2 in IDEF terms) within the system as well as the five different actors (five level 1 resources) that appear. The level 2 IDEF functions' use cases extend the level 1 functions' use cases. Selection sets selection policy and selects parts and sheets, Batching likewise sets batching policy and creates the batches. Assigning Machines and Set Scheduling Policy are use cases for Scheduling. As the Costing actor gives input to the other three actors for policy determination, it uses these cases. Likewise, as all four actors collaborate for the final cost evaluation, they all use this use case.

Use cases can extend the Set Selection Policy use case in the same way that the functions that appear in the level 3 IDEF diagram (Fig. 5) are extensions of higher level functions. The Set Part Selection Policy use case is extended by four use cases: Earliest Due Date, Special Order, Available Material, and Cost Effective. The Earliest Due Date use case is further refined to First Part and Largest Part, while the Cost Effective use case is extended to Cost Effective Part and Cost Effective Material. The level 3 use case extensions are conditional, as are the equivalent IDEF function extensions. A policy is decided upon once the present context or situation has been evaluated. Use case extensions for the other three policy determinations --- batching, scheduling, and costing --- are analogous to selection policy.

The dynamic model analyses object interactions for each use case. In the Set Selection Policy use case, the actor sets the policy. This policy is then reviewed by the Costing actor during cost evaluation, and the policy may be accepted or rejected, or a new policy may be suggested. Similar steps occur in the Set Batching and Set Scheduling Policy use cases. The similarity of the event traces and the state diagrams for the three use cases leads to the following conclusion; these three use cases can be derived from one abstract set or select policy use case. The evaluate costs use case, however, is derived from an abstract evaluation use case in that the costing actor reviews the policies and provides feedback. The concrete use cases specialise the abstract use cases, inheriting all of the features of the abstract case.

4.5 The Agent Oriented System
4.5.1 The Agents
Based on the simplifications and assumptions specified above, the agent oriented system for the case study consists of four agents for automated reasoning and collaboration. Agents are needed to set policies and formulate decisions: one each for setting selection, batching, and scheduling policies, and a costing agent for policy determination and cost evaluation. The agents Selection, Batching, Scheduling, and Costing replace some or all of the actor's behaviour in the use cases that produce control output. This determination is made on the basis of the use cases in Fig. 6 and the IDEF level 2 diagrams (Fig. 4).

4.5.2 Agent Goals and Plans
Once the agents have been identified, their goals and plans must be established, along with their beliefs. The Selection agent will be discussed as an illustrative example.

The goal of the Selection agent is to set selection policy; if the agent has meta-level knowledge this may be represented by strategic goals. The agent's plans arise from the IDEF level 3 diagram in Fig. 5 and the use case extensions. Each of these maps directly to a plan, as shown in Fig. 7. The agent then has six plans that it can draw on to achieve the top goal of setting part selection policy: earliest due date - first part, earliest due date - largest part, available material, special order, cost effective part and cost effective material.

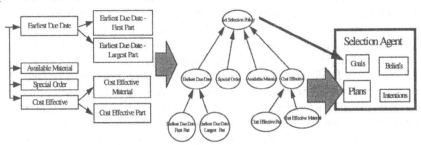

Fig. 7. The Selection Agent's Plans

An illustration of a plan is given in Fig. 8. These plans are written in dMARS (a hybrid C++/ dMARS implementation is under development). A plan is instantiated from the posting of a goal by the agent. The process proceeds down the plan from start to end following the links (P1, P2 etc.). If a link fails then the agent will proceed to an alternative route, however if no alternative route is given then the entire plan fails. If the process reaches the end, then the plan was successful.

Fig. 8 is a plan to find the first part with the earliest due date (Plan 1). Plan 1A is a sub-plan of plan 1 which is instantiated from a sub-goal to test for the earliest due date. It returns the earliest due date of all parts from the belief's database.

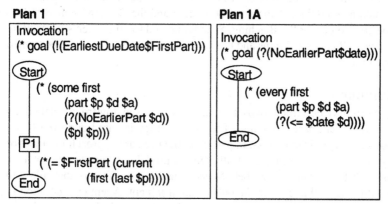

Fig. 8. Earliest Due Date - First Part Plan

4.5.3 Agent Communication

The Selection agent does not work alone; it collaborates with the Costing agent in the cost effective plans. This is evident from the control output from Costing that enters the IDEF level 3 diagram in Fig. 5. A plan with collaboration is not intrinsically different from a plan without it. All plans have a set of tasks that are meant to achieve the goal; plans with collaboration follow a communication protocol for these tasks. A simple example is in Fig. 9, where only an "ask" and "reply" protocol is given. Plan 2 demonstrates the Selection agent requesting the cost of a material from the Costing agent. Plan 3 is the Costing agent replying to the Selection agent with the cost of the material.

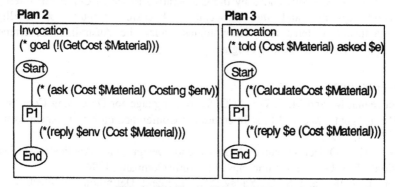

Fig. 9. Get Cost of Material Plan

Similar plans are needed for the Batching agent to collaborate with the Costing agent during its change batch plan, and for the Scheduling agent in the change assignment plan.

Discussion in the above sections has focused on the Selection agent, with some information about the collaboration between it and the Costing agent. Similar steps can be carried out for the Batching, Scheduling and Costing agents to result in IDEF$_0$ level 2 and 3 diagrams and use case diagrams which can then be transformed to agent plans.

5 Summary

This paper has described steps that have been taken toward a methodology for developing agent based systems. The methodology is based upon IDEF workflow and object oriented techniques, in particular the use case approach to object oriented software engineering (OOSE). With this methodology, agents can be identified, along with their plans, goals, beliefs and sensors and effectors that allow them to deal with objects that exist in the outside environment. Agent collaboration has also been addressed, and the methodology for the analysis and design of the collaboration is based on use cases and use case abstraction.

The agent oriented system for the case study is presently under development in dMARS and C++. Whereas the initial system will utilise passive objects as beliefs, an integrated agent and object design and implementation will be carried out. For design, the role of patterns, in particular the active object pattern (Lavender and Schmidt, 1995), the sensor and broker patterns (Coplien and Schmidt, 1995), and the mediator pattern (Gamma *et al.*, 1994), is under investigation. When this work is finished, a more complete statement of the agent development methodology will be formulated.

6 Acknowledgments

This project was partly funded by the Cooperative Research Centre for Intelligent Decision Systems and the Cooperative Research Centre for Intelligent Manufacturing Systems and Technologies under the Australian Government's Cooperative Research Centres Program.

7 References

Barbuceanu, M. and M. S. Fox, "COOL: A Language for Describing Coordination in Multi-Agent Systems," First International Conference on Multi- Agent Systems, pp. 17 - 24, 1995.

Booch, G., "Object Oriented Analysis and Design with Applications, Second Edition", The Benjamin/ Cummings Publishing Company, 1994.

Bravoco, R. R., and S. B. Yadav, "Requirement Definition Architecture - An Overview", Computers in Industry, Vol. 6, pp. 237- 251, 1985a.

Bravoco, R. R., and S. B. Yadav, "A Methodology to Model the Functional Structure of an Organization", Computers in Industry, Vol. 6, pp. 345- 361, 1985b.

Brenton, A., "The dMARS Plan Language Reference Manual", Australian Artificial Intelligence Institute (AAII), 1995.

Carrol, J.M., *Scenario-Based Design - Envisioning Work and Technology in System Development*, Wiley, 1995.

Cohen, D., and G. Ringwood, "Decision Table Languages and Guarded Definite Cleauses: Old Wine in New Bottles", Queen Mary and Westfield College, Dept of Computer Science Report, 1995.

Coplien, J.O., and D.C. Schmidt, *Pattern Languages of Program Design*, Addison Wesley, 1995.

Dunskus, B. V., D. L. Grecu, D. C. Brown, and I. Berker, "Using Single Functions Agents to Investigate Conflict", AI in Engineering Design and Manufacturing, special issue in Conflict Management in Multi-Agent Systems, to appear, 1995.

Finin, T., J. Weber, et al, "Specification of the KQML Agent- Communcation Language", The DARPA Knowledge Sharing Initiative External Interfaces Working Group, February, 1993.

Gamma, E., Helm, R., Johnson, R., and Vlissides, J., *Design Patterns - Elements of Reusable Object-Oriented Software*, Addison Wesley, 1995.

Genesereth, M., R. E. Fikes, et al, "Knowledge Interchange Format, Version 3.0, Reference Manual," Report Logic- 92- 1, Computer Science Department, Stanford University, 1992.

Georgeff, M. P., and F. Ingrand, "Research on Procedural Reasoning Systems", SRI International, 1990.

Gruber, T. R, "Ontolingua: A Mechanism to Support Portable Ontologies", Reference Manual, June, 1992.

Huntbach, M. M., N. R. Jennings, and G. A. Ringwood, "How Agents Do It In Stream Logic Programming", Proc. First International Conference on Multi- Agent Systems, pp. 177- 184, 1995

Jacobsen, I., *Object- Oriented Software Engineering: A Use Case Driven Approach*, Addison Wesley, 1992.

Klein, M., "Business Process (Re-)Engineering: Methodologies and Multi- Agent Technologies," Tutorial at First International Conference on Multi-agent Systems, 1995.

Kuwabara, K., T. Ishida, and N. Osato, "AgenTalk: Describing Multiagent Coordination Protocols with Inheritance," submitted to Tools for Artificial Intelligence Conference, 1995.

Lavender, R.G., and D.C. Schmidt, "Active Object: an Object Behavioural Pattern for Concurrent Programming", Pattern Languages of Programming Conference, Illinois, 1995.

Rao, A., private conversation, 1995.

Rumbaugh, J., M. Blaha, W. Premerlani, F. Eddy, and W. Lorensen, *Object Oriented Modeling and Design*, Prentice Hall, 1991.

Wooldridge, M. J., and N. R. Jennings, "Agent Theories, Architectures, and Languages", Tutorial, First International Conference on Multi- Agent Systems, 1995.

Plans and the Revision of Intentions

Wayne Wobcke
Knowledge Systems Group
Basser Department of Computer Science
University of Sydney, Sydney NSW 2006
Australia

Abstract

Although various theories of intention have been proposed, it is not obvious whether any of them can be realized in a computer implementation. Conversely, although there are many multi-agent systems, it is not clear which theories of intention they embody. In this paper, we present an initial step towards provably realizing a theory of intention in a computer system. The theory is a simplified version of Cohen and Levesque's theory based on situation semantics. We present a logic of belief and intention that is sound and complete with respect to our semantics and show that the standard logical puzzles concerning intention are handled correctly within the framework. The implementation is based on a belief revision system operating under the principle of minimal change of entrenchment. The main insight behind our approach is that persistence is not a defining property of intention, but rather is a consequence of the application of the principle of minimal change to intentions. A feature of the approach is the separation between the logic used by the agent and the dynamical properties of the agent's mental states. As a result, it is possible to define a simple rational agent whose intentions persist but which does not believe that its intentions persist. We show how linear hierarchical plans specified as ordered sets of beliefs and intentions can be represented and executed by a rational agent with the use of a simple interpreter.

1. Introduction

Motivated to varying degrees by Bratman (1987), there has been considerable interest in modelling rational agency in Artificial Intelligence using theories of intention. The work has two strands: a theoretical one, exemplified by Cohen and Levesque (1990b), Rao and Georgeff (1991a) and Konolige and Pollack (1993), and a practical one, typified by the application of decision theory to what, following Shoham (1993), might be called agent-oriented systems. The theoretical work stems in part from attempts to model a Gricean approach to meaning and pragmatics in natural language processing, Cohen and Perrault (1979), based on representing a speaker's conversational plans, Pollack (1990). The practical work may be said to arise from early work in reactive planners and multi-agent systems, although the incorporation of ideas on intention has occurred more recently, e.g. in the PRS system, Georgeff and Lansky (1987). Russell and Wefald (1991) provide an extensive analysis of the computational issues involved.

While some researchers, especially Rao and Georgeff (1992), have attempted to relate a theory of intention to aspects of its implementation (using PRS in their case), the two strands of research are not as connected as they might be. The following questions arise. First, given a theory of intention, can it be shown how a computer system realizing the theory can be implemented? Are realizations of the theory possible at all? Second, given an implementation, can it be proven what theory of intention the system obeys, and if so, whether it is a rational theory of intention? Answering these questions is further complicated, if as Shoham (1993) suggests, Dennett's (1987) intentional stance is adopted, wherein beliefs, desires and intentions are not explicitly part of the representation used by the system, but ascribed from the outside.

These questions motivate the work described in this paper. The overall research project is to develop theories of intention and implementations of agent-oriented systems in such a way that it can be proven that the theories of intention are realized in the systems. In this paper, we present an initial attempt at solving this problem, with a theory of intention based on a simpli-fied form of Cohen and Levesque's theory, and an implementation based on a belief revision system we have developed, Dixon and Wobcke (1993). The logic is a modal logic with explicit belief and intention operators, and the belief revision system operates using the principle of minimal change of entrenchment. The correspondence between theory and system is possible because intentions and beliefs are represented explicitly in the system, and so the philosophical basis for this work is closer to Fodor's (1975) language of thought than to Dennett's intentional stance. The main feature of the approach is that the persistence of intention is a property of the system that holds under certain assumptions, rather than being stipulated as part of the defini-tion of intention as in Cohen and Levesque's work. In our system, the persistence of intention is analogous to the persistence of belief as captured by Gärdenfors' (1988) notion of minimal change, and thus at the core of this work is an attempt to connect research on belief revision to that on planning, plan recognition and rational agency.

The organization of the paper is as follows. First, a brief summary of current theories of intention and agent-oriented systems is given with a view to evaluating their strengths and weaknesses for showing their correspondence. Second, a semantics for intention and belief is given which is based on situation semantics and belief revision, and a modal logic of belief and intention, sound and complete with respect to the semantics, is provided. The static properties of the logic are discussed with reference to the standard logical puzzles of intention, and the dynamics of the theory are given as a generalization of the AGM belief revision paradigm incorporating minimal change of entrenchment. This means that the persistence of intention is guaranteed by the principle of minimal change of theories applied to formulae representing intentions. Finally, we indicate how linear, hierarchical plans may be represented in the system using entrenchments over the logical language of beliefs and intentions, and define a simple interpreter capable of executing plans specified in this way. This goes one step towards realiz-ing rational agents whose plans are represented as complex mental attitudes, Pollack (1990).

2. Theories of Intention and Agent-Oriented Systems

2.1. Theories of Intention

A number of theories of intention have been proposed recently, e.g. Cohen and Levesque (1990b), Rao and Georgeff (1991a) and Konolige and Pollack (1993), and it is impossible here to give an extensive summary and critique of them all. Below, the main points of each theory are summarized and each theory is evaluated from the point of view of proving whether a sys-tem has intentions according to the theory.

Cohen and Levesque (1990b) provide a definition of what is to count as an intention using the notion of a persistent goal, or P-GOAL. The intuition is that a P-GOAL is a future goal of the agent that persists in the sense that the agent will retain the goal as long as the agent does not believe (i) that the goal has been achieved and (ii) that the goal will never be achieved. Persistence is thus a defining property of an intention. Moreover, for an agent to have p as a P-

GOAL at time t, the agent must believe $\sim p$ at time t. It is also possible to have P-R-GOALs, or P-GOALs relativized to other conditions. The semantics for belief and goal are given in terms of sequences of possible worlds, and the whole formalism rests on a foundation of dynamic logic.

A number of technical problems with the definitions proposed by Cohen and Levesque are noted in Rao and Georgeff (1991a) and Singh (1992). However, there are two more severe problems from the point of view of proving that a system realizes the theory. The first can be summarized in the statement that the theory of intention is static, rather than dynamic. This is because an agent's mental state is relative to points in time, but there is no connection between an agent's mental state at time t and at time $t + 1$. That is, the theory deems an agent rational if *for each time point* t, the possible future courses of events from the agent's point of view at time t ensure the persistence of the goals the agent has at time t. It is quite possible for an agent to completely change its mental state from time t to time $t + 1$; the agent remains rational so long as from the point of view of the agent at the time $t + 1$, the goals it has at that time point are persistent. Thus this definition has much in common with Bratman's ahistorical definition of agent rationality. The second problem is that even with the static theory of intention, P-GOALs do not actually persist as P-GOALs! More precisely, a P-GOAL p is a goal that, from the agent's point of view at time t, persists into the future as viewed from time t *as a goal*. However, it does not follow that the formula P-GOAL(p) holds at all of these future times. To see this, note that it is a requirement for P-GOAL(p) at some time that the agent believe $\sim p$ at that time. The agent may see itself as becoming agnostic with respect to p at some future time and hence losing the P-GOAL p at that time, while maintaining the goal p and hence the persistence of the goal p. Arguably, Cohen and Levesque's condition of believing $\sim p$ is too strong; however, as noted by Singh (1992), replacing it by the weaker non-belief in p destroys some of the desired properties of intention.

The theory of intention presented by Rao and Georgeff (1991a) is similar to Cohen and Levesque's but differs in allowing an explicit intention operator whose behaviour can be constrained axiomatically. Rao and Georgeff also argue for a branching-time semantics of intention. In Rao and Georgeff (1991b), it is shown how the logic correctly captures Bratman's asymmetry thesis (that one can intend p but not believe one will do p) and side-effect problem (one can intend p but not all of the known consequences of p). Apart from these differences, the theory of Rao and Georgeff is, like Cohen and Levesque's, a static one, although it might be expected that an instantiation of an agent using PRS would realize one of the possible paths out of the set of possible futures indicated in the time branches (thus providing a dynamic aspect to the theory). However, the persistence of a PRS agent's mental state through time is not discussed, and an agent is presumably deemed rational if, from the point of view of each time point, all possible futures from that time point meet the requirements for persistence of intentions. Since an intention is a property that holds in all the chosen 'intention-worlds', the theory does not suffer from the second weakness of Cohen and Levesque's theory: with respect to time t, the formula INTEND(p) does hold at all points in all futures which lead to p being fulfilled. However, the logic that is adopted is KD45 for belief and KD for intention, along with various axioms for the relationship between belief and intention. The presence of the (K) axiom for intention means that the conjunction problem is not averted: Bratman's video-game example shows that one can intend to hit each of two targets but not intend (in fact, believe that it is impossible) to hit both targets, but this intentional state is impossible to represent using Rao and Georgeff's logic.

The final theory to be discussed here is the 'representationalist' theory of Konolige and Pollack (1993). In this theory, a semantics of intention based on scenarios, or sets of possible worlds, is provided, and the conjunction problem is averted because the (K) axiom for intention is no longer valid. However, the definitions are too strict: on Konolige and Pollack's theory, an agent can intend p without even intending logical consequences of p such as $p \vee q$. It is strange that an agent can desire a world within a scenario satisfying p, yet not desire the very same world when it occurs within a scenario satisfying $p \vee q$. While the system E of Chellas (1980) is used by Konolige and Pollack, the use of the system M would also avert the conjunction problem (the difference between E and M is that if an agent intends p, with E the agent need only intend propositions logically *equivalent* to p, while with M the agent must intend all *monotonic* consequences of p). Perhaps what motivates Konolige and Pollack's choice of E rather than M is the desire to distinguish different possible behaviours of an agent which intends p; one agent might adopt q on giving up p while another might give up $p \vee q$ on giving up p. It is proposed here that these two agents be distinguished not by the contents of their mental states, but by the dynamics of their mental states.

2.2. Agent-Oriented Systems

Again, it is impossible to survey here all work in the application of ideas from theories of intention to the development of multi-agent systems or reactive planning systems. We are interested in how difficult it is to show which theory of intention a given system realizes, and we focus on the paradigm of agent-oriented systems as defined in Shoham (1993). An agent-oriented system is a specialization of an object-oriented system where the objects are agents, the data structures are called beliefs, goals and intentions (and possibly other attitudes) of the agents, and the messages correspond to speech acts such as requesting and informing. The basic processing loop stems from work in discrete event simulation: at each time point, an agent first updates its mental state and then executes any prior commitments for that time.

The major problem within the framework is in determining whether an agent's theory of belief and intention, etc. is the correct one. More precisely, on the one hand, an agent has beliefs, goals and intentions in the internal representation of its 'mental state', and on the other, is ascribed beliefs, goals and intentions from the outside according to Dennett's intentional stance. The problem of providing a 'demonstration that the component of the machine obeys the theory' is recognized by Shoham (1993, p.54) but never addressed since he never provides a theory of intention. This problem is considered in a multi-agent setting by Jennings (1995), but here the need for agents to communicate their intentions to one another favours the explicit representation of intentions.

To illustrate the problem, showing that AGENT-0 satisfies Cohen and Levesque's theory of intention would require determining the conditions under which a commitment can be given up. In AGENT-0, an agent gives up a commitment when (i) the agent no longer believes it is capable of performing the act or (ii) an agent receives a request to remove the commitment. This second condition clearly violates the property of persistence (which, recall, Cohen and Levesque (1990b) take to be a *defining* property of intention) and hence AGENT-0 simply does not have intentions in the Cohen and Levesque sense. The point of the discussion is not to criticize AGENT-0, but rather to show that determining the theory of intention by which a system operates is a non-trivial task.

3. The Dynamics of Belief and Intention

It is proposed to model the dynamics of a rational agent's intentional states by adopting the AGM paradigm for theory change. In this approach, an agent's intentional state is represented as a theory over a suitable base logic of intention and belief: such a theory is analogous to a plan in the sense of Pollack (1990). The resulting dynamical theory of intentional states can then be realized using an implementation of a belief revision system such as that described in Dixon and Wobcke (1993). This system is based on a generalization of the principle of minimal change to apply to the entrenchments of beliefs in addition to the beliefs themselves, allowing iterations of belief change operations, and is described more fully in Wobcke (1995). The intuition guiding our approach is that the persistence of intention is nothing but the effect of the principle of minimal change as applied to intentions. Thus, in contrast to Cohen and Levesque (1990b), the persistence of intention is not a defining property of intention, but is a special case of the principle of minimal change understood as a general principle of rationality. Similar considerations seem to motivate the work of Bell (1995) in which persistence follows from the properties of chronologically minimized models, Shoham (1988).

We now develop the technical foundations of the approach more precisely. In section 3.1, we define a suitable logical language in which the beliefs and intentions of an agent can be expressed. For this logic, we take a modal logic of belief and intention which is sound and complete with respect to a semantics of belief and intention based on situation semantics. Then in section 3.2, we review our theory change operations and how they may be represented and computed using E-bases. This separation of the static logic of belief and intention from the dynamics of intentional states makes the resulting formalism simpler than that described by Asher and Koons (1993), who try to capture both aspects of the theory in one formal system based on dynamic logic.

3.1. A Logic of Belief and Intention

The semantics for belief and intention we propose is a simplified version of that provided by Cohen and Levesque (1990b), based on situations or partial states of affairs, Barwise and Perry (1983): such a theory was anticipated by Cohen and Levesque (1990a, p.67). The theory is also closely related to that given in Konolige and Pollack (1993). For simplicity, only the case of non-nested intentions is considered here (although an agent can introspect on its intentions). Also, intentions are assumed to be non-specific temporally, i.e. an agent can intend to go to the dentist, but not intend to go to the dentist on Tuesday at 9.00 a.m.

The semantics of intention is based on sequences of situations. The intuition is as follows. A sequence of situations can be viewed as a possible future (starting at the current time point) containing the actions that the agent sees itself as performing intentionally. An agent's intentional state (at some point in time) is modelled as a set of such possible futures. Our semantics captures (only) the future-directedness of intention; in this simplified setting, an agent intends p if p holds in at least one future situation in every sequence of the agent's possible future intentional actions. Because the situations in the possible futures are supposed to satisfy only those actions the agent considers itself as endeavouring to execute intentionally, we are formally relating the future-directedness of an intention with the future intentional performance (more precisely, attempted performance) of the corresponding action.

Definition. An *intentional state* σ is defined to be a nonempty set σ_i of (partially defined) sequences of pairs $<B_i(t), I_i(t)>$ such that (i) each B_i and I_i are defined for $t = 0$ (the current time point), (ii) each $B_i(t)$ overlaps $I_i(t)$, and (iii) the future parts of the intention sequences are either jointly coherent or all empty. The subsidiary definitions are as follows.

Definition. A *situation* is a non-empty set of models of a given, fixed propositional language.

Definition. Two situations *overlap* if they have a non-empty intersection.

Definition. The *future part* of an intention sequence $I(t)$ is the subsequence of I over which $t > 0$. Note that this subsequence is empty if I is undefined for all such t.

Definition. A number of intention sequences are *jointly coherent* if the situations contained in them (taken together) form a coherent set of situations.

Definition. A set of situations is *coherent* if its elements have a non-empty intersection.

The intuitive meaning of two situations overlapping is that the sets of facts satisfied by the situations cannot be jointly inconsistent; coherence is the generalization of overlapping to two or more situations. Thus condition (ii) says that an agent's intentions must be consistent with its future beliefs, while condition (iii) says that the intended actions of an agent must be consistent as a whole.

In the following truth definitions, ϕ is restricted to be a propositional formula while α and β are combinations of belief formulae and intention formulae using the propositional connectives and the modal operator B (but not the operator I). A *belief formula* (*intention formula*) is a formula of the form $B\phi$ ($I\phi$) where ϕ is a propositional formula. Thus the restriction on the language is essentially that an intention formula cannot occur within the scope of an I operator.

$$\sigma_i(t) \vDash B\phi \quad \text{if all models in } B_i(t) \text{ satisfy } \phi$$
$$\sigma_i(t) \vDash I\phi \quad \text{if for some } t' > t, \text{ all models in } I_i(t') \text{ satisfy } \phi$$
$$\sigma_i(t) \vDash B\alpha \quad \text{if } \sigma_i(t) \vDash \alpha$$
$$\sigma_i(t) \vDash \alpha \wedge \beta \quad \text{if } \sigma_i(t) \vDash \alpha \text{ and } \sigma_i(t) \vDash \beta$$
$$\sigma_i(t) \vDash \alpha \vee \beta \quad \text{if } \sigma_i(t) \vDash \alpha \text{ or } \sigma_i(t) \vDash \beta$$
$$\sigma_i(t) \vDash \sim\alpha \quad \text{if } \sigma_i(t) \nvDash \alpha$$
$$\sigma_i(t) \vDash \alpha \to \beta \quad \text{if } \sigma_i(t) \nvDash \alpha \text{ or } \sigma_i(t) \vDash \beta$$

Only the first three truth rules require any discussion. By the definition of a situation as a set of models, any situation satisfies all the propositional tautologies and the class of situations is thus characterized (but not by a one-one mapping) by the propositional theories over the given language. Thus by the first and third truth rules, the logic of belief is **KD45**. By the second truth rule, the logic of intention contains the system M of Chellas (1980), although note that nested intentions are not allowed. Because the intention sequences are jointly coherent, the logic of intention contains the (D) axiom. Also note that by the third truth rule, an agent has positive and negative introspection of its intentions as well as of its beliefs.

An intentional state σ satisfies α at time t if for all i, $\sigma_i(t) \vDash \alpha$. Thus the logic of belief and intention BI determined by the above semantics (the formulae holding at all intentional states) can be summarized as follows. Again, α and β are combinations of belief and intention formulae and ϕ and ψ are restricted to be non-modal formulae.

(PC) Propositional Calculus tautologies
(MP) From α and $\alpha \to \beta$ infer β
(K) $B(\alpha \to \beta) \to (B\alpha \to B\beta)$
(D) $B\alpha \to \sim B\sim\alpha$
(4) $B\alpha \to BB\alpha$
(5) $\sim B\alpha \to B\sim B\alpha$
(N) If $\vdash \alpha$ infer $B\alpha$
(DI) $I\phi \to \sim I\sim\phi$
(MI) If $\vdash \phi \to \psi$ infer $I\phi \to I\psi$
(I+) $I\phi \leftrightarrow BI\phi$
(I–) $\sim I\phi \leftrightarrow B\sim I\phi$

Theorem. The logic BI is sound and complete with respect to the class of intentional states.

To conclude this section, we evaluate the logic BI and the semantics provided above as a logic of intention. Note that because of our restriction on the objects of intention to non-modal propositions, some of the desired properties of intention hold but cannot be expressed in BI. An example of such a property is what Rao and Georgeff (1991b) call weak realism, the property that if an agent intends p, the agent must believe p to be a possible future action. Since each intention sequence is meant to model one possibility for the agent's future actions, weak realism will follow if the following truth rule is adopted: $\sigma(t) \models \phi$ if ϕ holds in one of the $\sigma_i(t')$ for some $t' > t$. This would result in the logic S5 for epistemic necessity.

The use of KD45 as the logic of belief means that our agents are logically omniscient concerning their beliefs and intentions, although because the (K) axiom and necessitation rule are invalid for intention, our agents do not intend all the tautologies and do not intend the consequences of their intentions. Thus the side-effect problem is handled correctly: an agent can intend to go to the dentist yet not intend pain, even though it can believe it will be in pain. The fact that our intention sequences are required only to be jointly coherent means that the asymmetry problem is also avoided: if an agent intends p, it does not have to believe that p will occur. Similarly, what Rao and Georgeff (1991b) call the transference problem is avoided. The transference problem is that an agent should not be forced to intend everything that it thinks will happen anyway. This is avoided because the intention sequences are supposed to satisfy only those actions the agent sees itself as executing intentionally. Finally, nothing in the logic guarantees the persistence of intention in time: this will follow from the dynamics of intention.

3.2. Belief and Intention Revision

We are now in a position to define the dynamics of mental states. Here, the term 'mental state' is used to refer to a theory of the logic BI: it is thus the representational analogue of an intentional state – while an agent is *in* an intentional state, it *has* a mental state. The mental state of an agent at time t corresponds to an intentional state composed of a set of sequences all of whose initial segments up to and including time t are identical (the initial segments represent the history of the agent). The dynamics of mental states is defined by adopting the approach to theory change developed by Alchourrón, Gärdenfors and Makinson (1985) together with the principle of minimal change of entrenchment as described in Dixon and Wobcke (1993) and Wobcke (1995). Applying a revision operation takes the mental state of the agent at time t to its mental state at time $t + 1$.

In the AGM paradigm, the *expansion* of the belief set K by a formula A, denoted K_A^+, represents the addition of A to K regardless of whether it is consistent with K. Writing Cn for the logical consequence operator, the postulates for expansion imply that $K_A^+ = Cn(K \cup \{A\})$. The *revision* of a belief set K by a formula A, denoted K_A^*, represents the change made to K in order to accept new information A. The following 'rationality postulates' were proposed as desirable properties for a belief revision operation to satisfy.

(K*1) K_A^* is a belief set
(K*2) $A \in K_A^*$
(K*3) $K_A^* \subseteq K_A^+$
(K*4) If $\sim A \notin K$ then $K_A^+ \subseteq K_A^*$
(K*5) $K_A^* = K_\perp$ only if $\vdash \sim A$
(K*6) If $\vdash A \leftrightarrow B$ then $K_A^* = K_B^*$
(K*7) $K_{A \wedge B}^* \subseteq (K_A^*)_B^+$
(K*8) If $\sim B \notin K_A^*$ then $(K_A^*)_B^+ \subseteq K_{A \wedge B}^*$

In the above definitions, $K_\perp$ is the inconsistent set containing all formulae of the language.

Given a collection of belief sets over some logic, there are many revision functions satisfying the AGM postulates. To specify a particular revision function, Gärdenfors and Makinson (1988) use a total ordering on the formulae of the language, called an *epistemic entrenchment*. The conditions for an ordering ≤ to be an epistemic entrenchment are as follows.

(EE1) If $A \leq B$ and $B \leq C$ then $A \leq C$

(EE2) If $A \vdash B$ then $A \leq B$

(EE3) For any A and B, $A \leq A \wedge B$ or $B \leq A \wedge B$

(EE4) When $K \neq K_\perp$, $A \notin K$ iff $A \leq B$ for all B

(EE5) If $B \leq A$ for all B, then $\vdash A$

A unique revision operation is determined by an epistemic entrenchment using the (C*) condition.

(C*) $B \in K_A^*$ iff either $\vdash \sim A$ or $\sim A < A \rightarrow B$

Epistemic entrenchments themselves are not suitable for computation because too much information (the entrenchment relation between *every* pair of formulae) has to be provided. Rott (1991) investigated the use of so-called E-bases for specifying entrenchments. The formulation we use is the special case of an E-base in which the base is well-ordered, as considered by Williams (1992). Define a *ranked E-base* to be a theory base together with an assignment to each element of the base a natural number known as its *rank*. The given ranking on the base formulae is understood to extend in a unique way to the most conservative entrenchment compatible with the ranking, as described in Wobcke (1995). Williams (1992) shows that this entrenchment can be defined as follows. Suppose the E-base Γ is partitioned into subsets $\Gamma_1, \Gamma_2, \cdots$ where Γ_i is the set of formulae with rank i. The ranking on the base must satisfy the following condition.

(R) For all $A \in \Gamma$, $\{B \in \Gamma \mid rank(B) > rank(A)\} \nvdash A$

Let $\overline{\Gamma}_i = \overset{n}{\underset{j=i}{\cup}} \Gamma_j$. Intuitively, $\overline{\Gamma}_i$ is the set of beliefs in the base which are held with strength at least i. Then the rank of a belief A is the largest i such that A is a consequence of $\overline{\Gamma}_i$. A ranked E-base then generates an epistemic entrenchment by setting $A \leq B$ if and only if $rank(A) \leq rank(B)$ or B is a theorem.

A revision operation that captures the principle of minimal change of entrenchment can now be defined. Note that in defining such a revision, the rank of the added formula must be given, as there is no way to automatically determine the strength of a newly acquired belief. If $K_{A,\alpha}^*$ is the state resulting from a revision of K to accept A with rank α, the ranking $rank_{A,\alpha}^*$ on $K_{A,\alpha}^*$ is defined as follows.

$$rank_{A,\alpha}^*(B) = max(rank(B), min(rank(A \rightarrow B), \alpha)) \text{ if } B \in K_{A,\alpha}^*$$

Given a ranking on a theory base Γ representing a belief set K, Dixon and Wobcke (1993) show how to compute a theory base Δ and its ranking $rank_{A,\alpha}^*$ that represents the belief set $K_{A,\alpha}^*$ resulting from the revision of K to accept A with rank α. The base Δ consists of the formulae in Γ ranked higher than $\sim A$ (ranked as in Γ) together with the formulae $A \vee B$ such that $B \in \Gamma$, $\alpha < rank(B) \leq rank(\sim A)$ and $B \in K_{A,\alpha}^*$ (ranked as B is in Γ), pruned to remove redundancy. A formula B is redundant in the base Δ if it is derivable from the formulae in Δ with a higher rank. It is easy to verify that the ranking on Δ represents the ranking $rank_{A,\alpha}^*$ defined above.

With the above definitions, we can now realize the property of persistence for intentions using an implementation of an AGM revision system. We suppose the AGM system starts with a ranked E-base Γ over the logic BI. The system then is subject to successive revisions by formulae $B\phi_1, \cdots$, which are meant to correspond to beliefs the agent acquires from observing the world, e.g. after performing an action. Suppose a mental state contains the formula Ip. Then, provided the revisions are at appropriate ranks, the mental states resulting from the successive revisions will also contain Ip unless a formula $B\phi_i$ is accepted such that $\sim B\phi_i$ is entrenched the same as $B\phi_i \rightarrow Ip$ in the immediately preceding state. That is, the formula Ip persists as long as $\sim B\phi_i$, and so in some sense, the intention is conditional on $\sim B\phi_i$. We now give some examples of plans represented as mental states and show how persistence arises from the execution of a plan simulated using successive AGM revisions.

4. Plans as Complex Mental Attitudes

To illustrate how simple plans may be represented in a belief revision system as theories of BI and also to illustrate the persistence of intentions, we present some examples of the system in operation. The first example also illustrates how even simple mental states can be used to represent conditional intentions, which are difficult to represent in some planning formalisms. The second example shows more generally how linear hierarchical plans can be represented as complex mental states. The creation of intentions cannot be modelled in the system, so the initial state is assumed to represent the agent's initial plan.

In the first example, the initial scenario is as follows. The agent believes it has a toothache, and intends to go to the dentist. The dentist is in town. The agent believes it can get to town by train and by bus, and adopts an intention to go by train. The agent believes that if the train is delayed, it should take the bus, and that taking the train requires the purchase of a train ticket while taking the bus requires the purchase of a bus ticket. The agent believes it already has a bus ticket but does not believe it has a train ticket.

An appropriate BI theory and ranking for representing the initial information is as follows (there are many theories which would work just as well). Note that the more strongly held beliefs (such as believing that one must go to town if one intends to see the dentist and believes the dentist to be in town) have high rank, whereas beliefs which can change more easily (such as having a toothache) have low rank. The actions in the 'plan' correspond to the formulae p such that Ip is a consequence of the theory.

250: $Bin(dentist, town) \wedge Igo(dentist) \rightarrow Igo(town)$
200: $Igo(town) \rightarrow Igo(town, train)$ XOR $Igo(town, bus)$
150: $Bhas(toothache) \rightarrow Igo(dentist)$
120: $Bdelay(train) \rightarrow Igo(town, bus)$
100: $Igo(town, train) \wedge \sim Bhas(train-ticket) \rightarrow Ibuy(train-ticket)$
100: $Igo(town, bus) \wedge \sim Bhas(bus-ticket) \rightarrow Ibuy(bus-ticket)$
90: $Bdelay(train) \rightarrow Bhas(train-ticket)$
80: $Bhas(toothache)$
80: $Bhas(bus-ticket)$
70: $Bin(dentist, town)$
60: $Igo(town, train)$
50: $\sim Bhas(train-ticket)$

As a consequence, the agent intends to go to the dentist (rank 80), to go to town (rank 70) and to buy a train ticket (rank 50), and believes it doesn't intend to take the bus and that the train won't be delayed (rank 60).

There are a number of possible revisions that might be made at this point. For example, if the agent accepted that it didn't, after all, have a toothache, i.e. accepted $\sim Bhas(toothache)$, it would drop the intention to go to the dentist because $Bhas(toothache) \vee Igo(dentist)$ is ranked the same as B$has(toothache)$ (rank 80). Similarly, the agent would drop its intention to go to town by train because B$has(toothache)$ ∨ I$go(town, train)$ also has rank 80. If the agent were to accept that the dentist was not in town, the intention to go to town would be dropped but not the intention to go to the dentist.

Suppose that what actually happens is that the agent buys a train ticket. Assuming the action to be successful, this is modelled as a revision to accept the postcondition of the action, B$has(train-ticket)$ at rank 50. No other beliefs are changed and no intentions are dropped. Next suppose the agent learns that the train has been delayed, accepting B$delay(train)$, also at rank 50. The intention to go to town by train is dropped because B$delay(train)$ → I$go(town, train)$ has rank 60, the same as I$go(town, train)$. The mental state now contains B$delay(train)$ ∨ I$go(town, train)$ at rank 60, so that on giving up the belief that the train had been delayed, the intention to take the train would be recovered. Also, the agent forms the intention to take the bus to town (rank 50) but not to buy a bus ticket. Thus the original mental state can be said to contain a conditional intention, and the original plan a conditional action (taking the bus conditional on learning of the train's delay). The belief B$delay(train)$ → B$has(train-ticket)$ at rank 90 encodes the independence of the beliefs ~B$delay(train)$ and B$has(train-ticket)$, so the agent still believes it has a train ticket on learning of the delay.

The mental state of the agent after performing the two revisions is as follows.

> 250: B$in(dentist, town)$ ∧ I$go(dentist)$ → I$go(town)$
> 200: I$go(town)$ → I$go(town, train)$ XOR I$go(town, bus)$
> 150: B$has(toothache)$ → I$go(dentist)$
> 120: B$delay(train)$ → I$go(town, bus)$
> 100: I$go(town, train)$ ∧ ~B$has(train-ticket)$ → I$buy(train-ticket)$
> 100: I$go(town, bus)$ ∧ ~B$has(bus-ticket)$ → I$buy(bus-ticket)$
> 90: B$delay(train)$ → B$has(train-ticket)$
> 80: B$has(toothache)$
> 80: B$has(bus-ticket)$
> 70: B$in(dentist, town)$
> 60: B$delay(train)$ ∨ I$go(town, train)$
> 50: B$delay(train)$

Thus in this example, the intention to go to the dentist persists as long as the agent believes that it has a toothache, the intention to go to town persists as long as the agent intends to go to the dentist and believes the dentist is in town, and the intention to take the train persists as long as the the agent intends to go to town and believes the train is not delayed. In Cohen and Levesque's terminology, the intentions are P-R-GOALs rather than P-GOALs.

A question raised by the above discussion is how the ranks of the new beliefs are determined. As argued by Wobcke (1995), there can be no formal answer to this question because it concerns how the evidence for the new information relates to the evidence for existing beliefs: the chosen value crucially determines the future dynamical behaviour of the system. When a new belief is the postcondition of an action, the problem of determining its new rank presupposes a solution to the frame and ramification problems, although it is important to note that the principle of minimal change of entrenchment places real constraints on the belief revision procedure, which may not be compatible with (eventual?) solutions to these problems.

In the second example, we present an approach to representing any linear hierarchical plan as a mental state of beliefs and intentions, which clarifies how knowledge of a plan is represented using E-bases. Linear plans can be represented easily using our approach because a linear order on actions corresponds closely to the total pre-orders on formulae required by the

AGM approach. In a linear hierarchical plan, we assume there is one high-level action which decomposes into a linear sequence of actions, each of which may be further instantiated and decomposed, etc. The decompositions of some actions may overlap as long as each decomposition is linear. For an example, take the hierarchy of simple planning knowledge from a cooking domain, adapted from Kautz (1990), illustrated in Figure 1. In the figure, dashed arrows are used to indicate the abstraction relation between an instance of an action and its parent, and solid arrows to indicate the decomposition of an action into its component actions.

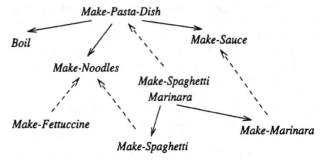

Figure 1. Cooking Hierarchy

The simple planning knowledge can be encoded in BI after making some assumptions about the meaning of the information represented in the hierarchy. For example, it is usually assumed that the postcondition of any decomposition or instantiation of an action entails the postcondition of the action, e.g. B$cooked(noodles)$ ∧ B$cooked(sauce)$ → B$cooked(pasta-dish)$ and B$cooked(fettuccine)$ → B$cooked(noodles)$. Similarly, the intention I$make(fettuccine)$ entails the intention I$make(noodles)$. One interpretation of the decomposition information is that a sufficient condition for intending a high level intention is that all its component actions are intended, e.g. I$make(spaghetti)$ ∧ I$make(marinara)$ entails I$make(spaghetti-marinara)$. Instantiation information implies that when an action must be instantiated to become executable, the intention to perform the parent action entails the intention to perform the instantiation, e.g. I$make(noodles)$ entails I$make(fettuccine)$ or I$make(spaghetti)$ but not both. Finally, an intention to perform an action should imply non-belief in its postcondition.

Suppose that the agent decides to make spaghetti marinara. We focus on the problem of constructing a mental state representing the agent's beliefs and intentions. First note that the general planning knowledge described above will be more entrenched than the intentions associated with this particular plan, which in this case are I$make(spaghetti-marinara)$, I$boil$, I$make(spaghetti)$, I$make(marinara)$, I$make(noodles)$, I$make(sauce)$ and I$make(pasta-dish)$. The main criterion for defining the ranking is the temporal ordering on the actions represented in the plan: intentions to perform later actions must persist longer and hence must be more entrenched than intentions to perform earlier actions. The ranking must also respect the conditions under which an intention should be dropped: i.e. if (i) its postcondition is achieved, or (ii) its parent intention is dropped. For example, consider the intention I$make(spaghetti)$ in the above plan. This should be dropped if (i) the postcondition B$cooked(spaghetti)$ is achieved, or (ii) the intention I$make(noodles)$ is dropped. In general, if p should be dropped on accepting a formula q then ~q should be ranked the same as ~q ∨ p, which means that p is not ranked more than ~q. Thus in this example, I$make(spaghetti)$ must be ranked no more than ~B$cooked(spaghetti)$ and I$make(noodles)$.

An appropriate BI theory and ranking representing this simple plan is as follows.

250: B$cooked(spaghetti{-}marinara)$ → B$cooked(pasta{-}dish)$
250: B$cooked(marinara)$ → B$cooked(sauce)$
250: B$cooked(fettuccine)$ ∨ B$cooked(spaghetti)$ → B$cooked(noodles)$
200: I$make(spaghetti{-}marinara)$ → I$make(pasta{-}dish)$
200: I$make(marinara)$ → I$make(sauce)$
200: I$make(fettuccine)$ ∨ I$make(spaghetti)$ → I$make(noodles)$
150: B$cooked(noodles)$ ∧ B$cooked(sauce)$ → B$cooked(pasta{-}dish)$
150: B$cooked(spaghetti)$ ∧ B$cooked(marinara)$ → B$cooked(spaghetti{-}marinara)$
100: I$boil$ ∧ I$make(noodles)$ ∧ I$make(sauce)$ → I$make(pasta{-}dish)$
100: I$make(spaghetti)$ ∧ I$make(marinara)$ → I$make(spaghetti{-}marinara)$
100: I$make(noodles)$ → I$make(fettuccine)$ XOR I$make(spaghetti)$
100: I$make(pasta{-}dish)$ → ~B$cooked(pasta{-}dish)$
100: I$make(sauce)$ → ~B$cooked(sauce)$
100: I$make(noodles)$ → ~B$cooked(noodles)$
 90: I$make(pasta{-}dish)$
 80: I$make(spaghetti{-}marinara)$
 80: I$make(marinara)$
 70: I$make(spaghetti)$
 60: I$boil$

Thus it can be seen that the ordering on the intention formulae reflects both the temporal ordering on the executable actions (the less entrenched the action, the earlier it should be performed) and the hierarchical relationship between actions in the plan and their decompositions (actions are at least as entrenched as their component actions). The intuitive understanding of entrenchment in belief revision is that less entrenched beliefs are easier to give up than more entrenched beliefs. Here, a possible intuition is that the agent wants to give up its intentions by replacing them with beliefs that their goals have been achieved, and less entrenched intentions are preferable to give up than more entrenched intentions – an intention is given up if the agent satisfies it by executing the corresponding action. An intention to perform a later action must persist longer and hence must be more entrenched than an intention to perform an earlier action. Similarly, an intention to perform a component action need persist only as long as the parent intention persists.

5. A Simple Agent Interpreter

We have shown how to represent linear hierarchical plans using an explicit language of beliefs and intentions, but two important questions remain. First, exactly how are intentions fulfilled and plans executed by a rational agent using this representation of plans? That is, how can a belief revision system be used to realize a planning agent? Second, if a rational agent uses formulae in a modal logic as explicit representations of its intentions, how can we justify the application of the term 'intention' to such mental entities? To address both of these questions, we define a simple agent interpreter whose plans are represented as epistemically entrenched intentions and beliefs. The dynamic behaviour of the interpreter (assumed connected to hardware able to perform individual actions) shows how complex plans are executed. The justification for ascribing intentions to an agent is that the behaviour of the interpreter as determined by the intention formulae satisfies Bratman's (1987) three functional roles of intention.

Our interpreter is a specialization of the following (simplified) BDI-interpreter defined by Rao and Georgeff (1992), based on PRS, Georgeff and Lansky (1987).

BDI-interpreter
initialize-state();
do
 options := option-generator(event-queue, B, G, I);
 selected-options := deliberate(options, B, G, I);
 update-intentions(selected-options, I);
 execute(I);
 get-new-external-events();
 drop-successful-attitudes(B, G, I);
 drop-impossible-attitudes(B, G, I)
until quit

A rational agent using this interpreter is reminiscent of a discrete event simulation system (although here the agent executes actions, rather than simulates execution). At each time cycle, the agent has a number of beliefs (B), goals (G) and intentions (I), and there are also a number of actions the agent can perform, represented in an event queue. Some of these actions are selected for possible execution, then these possibilities are further reduced by 'deliberation' (perhaps using some cost/benefit analysis of the options in relation to the current set of beliefs, goals and intentions). After execution, new facts are observed and the system state updated. Successfully fulfilled intentions and those now considered impossible to fulfil are dropped before the next cycle commences.

With a belief revision system using an explicit representation of plans, the system state consists of a base for a BI theory together with a ranking on the formulae of the base. The 'options' are represented by the formulae Ip in the agent's mental state and the 'selected options' are those formulae Ip that have the lowest rank. These formulae may correspond to actions that are directly executable, in which case one or more of them is executed, but if not, further intentions need to be created to fulfil them (analogous to specializing or decomposing a high-level action in a planner). There is no need to have the explicit steps of dropping fulfilled and unachievable intentions because this is supposed to follow from the operation of the revision algorithm on the agent's mental state.

BDI-interpreter
initialize-state(Γ);
do
 options := {Ip | $Ip \in \Gamma$};
 selected-options := lowest-ranked(options);
 if executable(selected-options)
  then execute(selected-options)
  else create-intentions(selected-options);
 get-new-external-observations(q);
 Γ := revise(Γ, Bq)
until quit

The above interpreter is limited to the extent that in the general BDI-interpreter, deliberation at each time cycle can reactively modify the entire plan of the agent, whereas we have assumed that the plan constructed by the system is consistent, so have not (yet) incorporated replanning into the model.

It is the dynamic behaviour of the interpreter that guarantees that the formulae of the form Ip have the properties of intentions as characterized by Bratman (1987). In Bratman's theory, intentions are defined as mental attitudes which have the following functional roles: (i) intentions pose problems for deliberation, i.e. how to fulfil them, (ii) existing intentions constrain the adoption of further intentions, and (iii) intentions control conduct: an agent endeavours to fulfil its intentions. The first property comes from the behaviour of the interpreter in creating intentions to fulfil those intentions corresponding to non-executable actions. The second property is assumed to follow from the operation of the procedure that creates these subsidiary intentions. Finally, the third property comes from the direct connection between intention formulae and the actions executed by the interpreter. Note also that for Bratman, the stability of intention is not one of its characterizing functional roles, but rather underwrites properties (i) and (ii) in the sense that these properties both generate and rely on stability. More precisely, the fact that existing intentions constrain further intentions generates stability, and the fact that intentions are stable ensures that deliberation can produce complex plans that can effectively achieve an agent's goals. Similarly, in our system, the persistence of intention is not definitional, but rather is captured by the principle of minimal change as applied to an agent's representational states.

6. Conclusion

Cohen and Levesque (1990b), Rao and Georgeff (1991a) and Konolige and Pollack (1993) all present logics of intention, but there is no indication of how these logics might be realized in a computer system. The agent-oriented programming paradigm is based on systems claimed to have intentions but there is no statement or justification of which properties the intentions have. In this paper, we have presented a simplified theory of intention with the consequence that the theory of intention can be realized in an AGM revision system, with the persistence of intention a special case of the principle of minimal change from belief revision. What makes the correspondence possible is the fact that agents have explicit knowledge of their intentions and beliefs. However, the main feature of our approach is the clear separation between the different components of the theory: (i) a logic of belief and intention forms the basis for an agent's representation of intentional states, and (ii) a theory of the dynamics of such states accounts for the persistence of intention. The agent does not have to have complete knowledge of the theory of intention, nor need it know or even believe in the persistence of its intentions: rationality ensues simply because the intentions do, in fact, persist. The formalism can be used to represent linear hierarchical plans and plans containing conditional actions.

Acknowledgements

This work was supported by an Australian Research Council Small Grant and the Symbolic Reasoning Systems Project of the Australian National University. Part of this work was carried out at the Center for the Study of Language and Information at Stanford University.

References

Alchourrón, C., Gärdenfors, P. & Makinson, D. (1985) 'On the Logic of Theory Change: Partial Meet Contraction and Revision Functions.' *Journal of Symbolic Logic*, **50**, 510-530.

Asher, N. & Koons, R. (1993) 'The Revision of Beliefs and Intentions in a Changing World.' *Proceedings of the 1993 Spring Symposium on Reasoning about Mental States: Formal Theories and Applications*, 1-9.

Barwise, J. & Perry, J. (1983) *Situations and Attitudes*. MIT Press, Cambridge, MA.

Bell, J. (1995) 'Changing Attitudes.' in Wooldridge, M.J. & Jennings, N.R. (Eds) *Intelligent Agents*. Springer-Verlag, Berlin.

Bratman, M.E. (1987) *Intention, Plans and Practical Reason*. Harvard University Press, Cambridge, MA.

Chellas, B.F. (1980) *Modal Logic: An Introduction.* Cambridge University Press, Cambridge.

Cohen, P.R. & Levesque, H.J. (1990a) 'Persistence, Intention, and Commitment.' in Cohen, P.R., Morgan, J. & Pollack, M.E. (Eds) *Intentions in Communication.* MIT Press, Cambridge, MA.

Cohen, P.R. & Levesque, H.J. (1990b) 'Intention is Choice with Commitment.' *Artificial Intelligence,* 42, 213-261.

Cohen, P.R. & Perrault, C.R. (1979) 'Elements of a Plan-Based Theory of Speech Acts.' *Cognitive Science,* 3, 177-212.

Dennett, D.C. (1987) *The Intentional Stance.* MIT Press, Cambridge, MA.

Dixon, S.E. & Wobcke, W.R. (1993) 'The Implementation of a First-Order Logic AGM Belief Revision System.' *Proceedings of the Fifth IEEE International Conference on Tools with Artificial Intelligence,* 40-47.

Fodor, J.A. (1975) *The Language of Thought.* Harvard University Press, Cambridge, MA.

Gärdenfors, P. (1988) *Knowledge in Flux.* MIT Press, Cambridge, MA.

Gärdenfors, P. & Makinson, D. (1988) 'Revisions of Knowledge Systems Using Epistemic Entrenchment.' *Proceedings of the Second Conference on Theoretical Aspects of Reasoning About Knowledge,* 83-95.

Georgeff, M.P. & Lansky, A.L. (1987) 'Reactive Reasoning and Planning.' *Proceedings of the Sixth National Conference on Artificial Intelligence (AAAI-87),* 677-682.

Jennings, N.R. (1995) 'Controlling Cooperative Problem Solving in Industrial Multi-agent Systems Using Joint Intentions.' *Artificial Intelligence,* 75, 195-240.

Kautz, H.A. (1990) 'A Circumscriptive Theory of Plan Recognition.' in Cohen, P.R., Morgan, J. & Pollack, M.E. (Eds) *Intentions in Communication.* MIT Press, Cambridge, MA.

Konolige, K. & Pollack, M.E. (1993) 'A Representationalist Theory of Intention.' *Proceedings of the Thirteenth International Joint Conference on Artificial Intelligence,* 390-395.

Pollack, M.E. (1990) 'Plans as Complex Mental Attitudes.' in Cohen, P.R., Morgan, J. & Pollack, M.E. (Eds) *Intentions in Communication.* MIT Press, Cambridge, MA.

Rao, A.S. & Georgeff, M.P. (1991a) 'Modeling Rational Agents within a BDI-Architecture.' *Proceedings of the Second International Conference on Principles of Knowledge Representation and Reasoning,* 473-484.

Rao, A.S. & Georgeff, M.P. (1991b) 'Asymmetry Thesis and Side-Effect Problems in Linear-Time and Branching-Time Intention Logics.' *Proceedings of the Twelfth International Joint Conference on Artificial Intelligence,* 498-504.

Rao, A.S. & Georgeff, M.P. (1992) 'An Abstract Architecture for Rational Agents.' *Proceedings of the Third International Conference on Principles of Knowledge Representation and Reasoning,* 439-449.

Rott, H. (1991) 'A Nonmonotonic Conditional Logic for Belief Revision.' in Fuhrmann, A. & Morreau, M. (Eds) *The Logic of Theory Change.* Springer-Verlag, Berlin.

Russell, S.J. & Wefald, E. (1991) *Do the Right Thing.* MIT Press, Cambridge, MA.

Shoham, Y. (1988) *Reasoning About Change.* MIT Press, Cambridge, MA.

Shoham, Y. (1993) 'Agent-Oriented Programming.' *Artificial Intelligence,* 60, 51-92.

Singh, M.P. (1992) 'A Critical Examination of the Cohen-Levesque Theory of Intentions.' *Proceedings of the Tenth European Conference on Artificial Intelligence,* 364-368.

Williams, M. (1992) 'Two Operators for Theory Base Change.' *Proceedings of the Fifth Australian Joint Conference on Artificial Intelligence,* 259-265.

Wobcke, W.R. (1995) 'Belief Revision, Conditional Logic and Nonmonotonic Reasoning.' *Notre Dame Journal of Formal Logic,* 36, 55-102.

A Formal View of Social Dependence Networks

Mark d'Inverno[1] and Michael Luck[2]

[1] School of Computer Science, University of Westminster, London, W1M 8JS, UK.
Email: dinverm@westminster.ac.uk
[2] Department of Computer Science, University of Warwick, Coventry, CV4 7AL, UK.
Email: mikeluck@dcs.warwick.ac.uk

Abstract. In response to the problems that have arisen regarding the terminology and concepts of agent-oriented systems, previous work has described a formal framework for understanding agency and autonomy. In particular, this work made the claim that the framework could serve as a vehicle for the precise presentation and evaluation of models and theories of multi-agent systems. We support this claim by outlining the framework and refining it through adding further levels of detail to formalise the concepts of *external descriptions* and *social dependence networks*. Social Dependence Networks are a valuable source of information about the relationships within a multi-agent world. They allow agents to reason about the resources and capabilities of others in order that they may enter into a negotiation to persuade these others to assist them in completing their tasks. By formalising social dependence networks within the framework we are able to identify deficiencies in the original characterisation of the networks and the external descriptions of agents within them. We address these deficiencies, and offer a modified view which removes much of the ambiguity and presents a stronger and more consistent formal model. In reformulating these networks in this way, we also present a case study which shows how the formal framework that has been previously developed can be applied to provide an environment in which we can describe and reason about theories and models of multi-agent systems.

1 Introduction

There is a growing recognition within the multi-agent system (MAS) community of the need to harmonise the efforts being made in different sub-fields and so derive a well-defined discipline of MAS [15]. Previously, we have developed a principled theory of agency and autonomy through the provision of a formal framework which defines these concepts and specifies the relationship between them [8]. This framework was an attempt to provide strong definitions, not only to be precise about the meaning of terms which often have an ambiguous interpretation, but also to serve as an environment in which theories and models of multi-agents systems can be presented, evaluated and developed. In this paper we illustrate how this can be done by adding detail to the framework to describe *social dependence networks* (SDN) [12]. Specifying SDNs formally in this way has allowed us to note inconsistencies and ambiguities in the work and suggest possibilities for its development as a useful mechanism for social agents.

As stated elsewhere[8], in the current work, we have adopted the specification language Z [14] for two major reasons. First, it provides modularity and abstraction and is sufficiently expressive to allow a consistent, unified and structured account of a computer system and its associated operations. Such structured specifications enable the description of systems at different levels of abstraction, with system complexity being added at successively lower levels. Second, we view our enterprise as that of building programs. Z schemas are particularly suitable in squaring the demands of formal modelling with the need for implementation by providing clear and unambiguous definitions of state and operations on state which provide a basis for program development. Thus our approach to formal specification is pragmatic — we need to be formal to be precise about the concepts we discuss, yet we want to remain directly connected to issues of implementation. Z provides just those qualities that are needed, and is increasingly being used for specifying frameworks and systems in AI [6, 3, 11] and related areas [4, 5].

The paper begins with a brief description of Social Dependence Networks [12], and continues with a very short outline of the agent hierarchy specified previously [8]. The next section extends and refines the specification of the agent hierarchy to specify SDNs formally. This allows us to evaluate and reason about these mechanisms in terms of our formal framework. We then develop and propose a refined model of dependence networks based on our notions of agency and autonomy. Lastly, we draw conclusions made from this case study of applying the framework we have developed.

2 External Descriptions and Dependence Networks

Dependence networks [12] are structures that form the basis of a computational model of Social Power Theory [1, 2]. They allow agents to reason about, and understand, the collective group of agents that make up the multi-agent world in which they operate. This section introduces dependence networks and external descriptions, data structures used to store information about other agents, based on the work reported by Sichman et al. [12].

External descriptions store information about other agents, and comprise a set of goals, actions, resources and plans for each such agent. The goals are those an agent wants to achieve, the actions are those an agent is able to perform, the resources are those over which an agent has control, and the plans are those available to the agent, but using actions and resources which are not necessarily available to the agent. This means that one agent may *depend* on another in terms of actions or resources in order for a plan to be executed.

An agent i is denoted by ag_i, and any such agent has a set of *external descriptions* of all of the other agents in the world, denoted by

$$Ext_{ag_i} \stackrel{\text{def}}{=} \bigcup_{j=1}^{n} Ext_{ag_i}(ag_j)$$
where

$$Ext_{ag_i}(ag_j) \stackrel{\text{def}}{\equiv} \{G_{ag_i}(ag_j), A_{ag_i}(ag_j), R_{ag_i}(ag_j), P_{ag_i}(ag_j)\}$$

such that

$G_{ag_i}(ag_j)$ is the set of goals,

$A_{ag_i}(ag_j)$ is the set of actions,

$R_{ag_i}(ag_j)$ is the set of resources, and

$P_{ag_i}(ag_j)$ is the set of plans

that agent i believes agent j has.

Notice that an agent has a model of itself as well as others. The authors adopt what they call the *hypothesis of external description compatibility* which states that any two agents will have precisely the same external description of any other agent. This is stated as follows.

$$Ext_{ag_i}(ag_i) = Ext_{ag_j}(ag_i) \wedge Ext_{ag_i}(ag_j) = Ext_{ag_j}(ag_j)$$

Now, $P_{ag_i}(ag_j, g_k)$ represents the *set* of plans that agent i believes that agent j has in order to achieve the goal g_k. Each plan within this set is given by $p_{ag_{i_l}}$, defined below:

$$p_{ag_{i_l}}(ag_j, g_k) \stackrel{\text{def}}{\equiv} \{g_k, R(p_{ag_{i_l}}(ag_j, g_k)), I(p_{ag_{i_l}}(ag_j, g_k))\}$$

where $R(p_{ag_{i_l}})$ represents the set of resources required for the plan and $I(p_{ag_{i_l}})$ is a *sequence* of instantiated actions used in this plan. Each instantiated action within a plan is defined by the action itself and the set of resources used in the instantiation of this action:

$$i_m(p_{ag_{i_l}}(ag_j, g_k)) \stackrel{\text{def}}{\equiv} \{a_m, R_{a_m}(p_{ag_{i_l}}(ag_j, g_k))\}$$

Note that this makes the definition of the resources of a plan redundant: if you know the resources required by each action within a plan, then you must also know the set of all the resources required by the plan.

3 An Overview of the Framework for Agency and Autonomy

Before we can attempt to reformulate the work described above in a broader formal framework, we must first provide an overview of that framework, specified in Z. Our basic component is an *entity* [10]. An entity consists of four constituents as follows: a set of attributes, which are perceivable qualities of the entity; a set of actions, which define the basic capabilities of the agent; a set of goals, which are the goals that can be ascribed to the entity which characterise its *agency*; and a set of internal non-derivable motivations which define an entity's *autonomy*.

```
┌─ Entity ──────────────────────────────
│ attributes : P Attribute
│ capableof : P Action
│ goals : P Goal
│ motivations : P Motivation
├───────────────────────────────────────
│ attributes ≠ {}
└───────────────────────────────────────
```

Using this schema we can define certain categories of entity. In particular, an object is any entity with a non-empty set of capabilities, an agent is any object with a non-empty set of goals, and an autonomous agent is any agent with a non-empty set of motivations.

```
┌─ Object ──────────────────────────────────────────────
│ Entity
│ ────────────────────────
│ capableof ≠ {}
└───────────────────────────────────────────────────────
```

```
┌─ Agent ───────────────────────────────────────────────
│ Object
│ ────────────────────────
│ goals ≠ {}
└───────────────────────────────────────────────────────
```

```
┌─ AutonomousAgent ─────────────────────────────────────
│ Agent
│ ────────────────────────
│ motivations ≠ {}
└───────────────────────────────────────────────────────
```

A full treatment of the framework which has subsequently been refined and developed in a number of ways can be found in [10]. A model of how goals are generated by motivated agents (which we take to be the defining quality of autonomy), and subsequently adopted by non-autonomous agents has been constructed [9]. We have also shown how certain social structures — cooperation between autonomous agents, and engagements of non-autonomous agents — arise as a result of such goal generation and adoption [7].

In addition, we can easily refine components within the framework to provide, for example, a high level specification of an autonomous planning agent. Consider the next schema which describes such a refinement. A planning agent is an agent with a set of plans associated with a set of goals. Each plan in the set is a possible means of bringing about the associated goal. Some subset of these goals are ones that the agent currently desires; it might have plans for a goal it does not currently desire. We define a *complete plan* to be a *sequence* of actions. (There are, certainly, other types of plan, but this will be sufficient for the presentation of SDNs in this paper.)

$$Plan == \text{seq } Action$$

```
┌─ PlanningAgent ───────────────────────────────────────
│ Agent
│ plans : P Plan
│ planforgoal : Goal ⇸ P Plan
│ ────────────────────────
│ goals ⊆ dom planforgoal
│ ⋃(ran planforgoal) = plans
└───────────────────────────────────────────────────────
```

The schema states that all the plans of an agent must be associated with a goal, although it may be that the set of plans associated with a goal is the empty set. It might also be that a plan brings about more than one goal of the planning agent.

4 Dependence Networks within the Formal Framework

By using this framework, specified in Z, as a basis for reformulating the model of SDNs, we can provide a clear and unambiguous formal model, and highlight some of the potential ambiguities which arise within the existing model. We take actions, goals and plans in the original model to be actions, goals and plans in the Z framework. We take a *resource* to mean some entity — an object, agent or autonomous agent.

4.1 External Descriptions

To deal with an *external description*, we must refine our definition of a simple planning agent by including three additional variables. The first, *ownedresources*, represents the set of resources which an agent *owns*. The second, *instsreq*, models the set of resources needed to instantiate an action within a plan. The third, redundant variable, *resourcesofplan*, is included for readability and records the total set of resources required by a plan.

There are two predicates in the lower part of the schema which relate the variables in the schema as follows: stripping the set of entities away from each instantiated action gives the original plan; and the resources of a plan are the union of each set of entities associated with each action of the plan.

```
┌─ ExternalDescription ─────────────────────────────────
│ PlanningAgent
│ ownedresources : ℙ Entity
│ instsreq : Plan ⇸ (seq (Action × ℙ Entity))
│ resourcesofplan : Plan ⇸ ℙ Entity
├────────────────────────────────────────────────────────
│ plans = mapset (mapseq first) (ran instsreq)
│ ∀ p : Plan • resourcesofplan p = ⋃(ran (mapseq second (instsreq p)))
└────────────────────────────────────────────────────────
```

Now, since every external description of an agent is the same, we can model the formalism very simply. An agent, A, has associated with it an external description which is precisely the model that every agent (including agent A) has of agent A (according to the hypothesis of external description compatibility).

```
┌─ World ────────────────────────────────────────────────
│ extdes : Agent ⇸ ExternalDescription
└────────────────────────────────────────────────────────
```

Then, according to the external description of some agent, i,
($extdes\ i$).*plans* is its set of plans,
($extdes\ i$).*capableof* is its set of actions,
($extdes\ i$).*ownedresources* is its set of resources and
($extdes\ i$).*goals* is its set of goals.

Discussion There are several difficulties that become apparent when the SDN model is reformulated in this way. First of all, the distinction between a resource and an agent is not clear. For example, is a benevolent agent, who will always adopt the goals of another, a resource or an agent? It seems that some arbitrary distinction, presumably, will have to be made. This distinction is important since the nature of a plan assumes that all of the *resources* of an action have already been identified, but the agents which could possibly perform some action have not. In this respect, a partial plan where the resources required (whatever they may be) have not yet been considered, cannot be represented.

It is also limiting in that two agents cannot perform the same action simultaneously. For example, the act of lifting a table might require two agents together performing a basic lift action. When reasoning about the multi-agent world in particular, where cooperation is likely to ensue, this seems a stringent restriction.

In addition, the notion of *ownership* in these external descriptions is not clear. We take it to mean that an agent *owns* another entity, if, for whatever the reason, that entity can be used for *any* action within its capabilities whenever the agent requires it. In other words, a resource in this formalism can be seen as a benevolent agent, adopting the goals of others' (to subsequently perform an action to satisfy those goals) whenever it can. Even then, however, there are further subtleties to consider. For example, many agents may *own* the same resource (such as a printer) but there is no mention of a shared resource. There may also be some *degree* of ownership in that my manager may always be able to use the printer before me, or some weaker notion of ownership like a desk in a shared office which can only be used by one of the occupants at a time. Clearly, a much richer notion of ownership is required. By contrast, the agent hierarchy allows us to be much clearer about the nature of these relationships which will depend on the type of entity and which goal dependence networks exist between the entities in the environment. If the entity required for some action is an object, then instantiating it as an agent is straightforward. If the entity is a non-autonomous agent, then the planning agent must reason about the nature of its agency further. (For example, can it share this agent? Can it persuade other agents that are currently engaging it to release it?) If the entity is an autonomous agent, then the planning agent will need to negotiate with the autonomous agent to persuade it to adopt its goal. More details of these social structures can be found in [9].

The *hypothesis of external description compatibility* ensures that any two agents will agree on the model of themselves and each other. Though the authors argue that there is no loss of generality, it is difficult to see how this can be so. A truly autonomous agent will have its own view of the world around it which may bear no relation to another agent's interpretation of its world. In general, we argue, any model of the world that an autonomous agent has of the world must be subjective. Certainly, a truly autonomous agent can never know the plans and goals of another agent; it may only infer them by evaluating the behaviour of the other agent. The authors themselves go some way along this path when they recognise in a later paper some of these difficulties [13], but they still require an agent to have complete (and correct) knowledge of other agents' plans, for example, which is untenable.

In Section 5, we will provide a formal specification which allows for concurrent actions in a plan, an important requirement of general purpose multi-agent systems. We do not arbitrarily distinguish agents from resources, but instead consider agents with different functionalities. In this way we can provide a clearer and more intuitive representation of the social structures in the world since a planning agent would have to consider merely the set of *agents* that are required in a plan. Some of these agents might be invoked directly, some might be shared with some other agent, and some might be autonomous agents requiring negotiation. These ideas are developed further in [7].

4.2 Definitions of Autonomy

Using external descriptions, Sichman et al. distinguish three different forms of autonomy. An agent is *a-autonomous* for a given goal according to a set of plans of another to bring about that goal if there is a plan in this set that achieves the goal, and every action in each plan belongs to the capabilities of the agent. An agent is *r-autonomous* for a given goal according to a set of plans of another to bring about that goal if there is a plan in this set that achieves the goal, and every resource required by the plan is owned by the agent. An agent is *s-autonomous* for a given goal if it is both *a-autonomous* and *r-autonomous*.

In the following schema, we define these three classes of autonomy using of a new relation, achieves. The predicate, achieves (a, g, ps), holds precisely when an agent, a, has goal, g, and the non-empty set of plans associated with g in order to achieve it, is ps.

Thus in the schema below, the first predicate states that an agent, a, is *a-autonomous* with respect to some set of plans, ps, if and only if there is some agent, c, with goal, g, and plans, ps, to achieve g such that some plan, p in ps, contains actions all in the capabilities of a. Similar predicate are specified for *r-autonomous* and *s-autonomous*. Finally, the achieves predicate is specified as defined above.

```
__ AutonomyRelations _____
  World
  aaut _, raut _, saut _, achieves _ : P(Agent × Goal × P Plan)
 _____
  ∀ a : Agent; g : Goal; ps : P Plan •
      aaut (a, g, ps) ⇔ (∃ c : Agent • achieves (c, g, ps)) ∧
                      (∃ p : ps • (ran p ⊆ (extdes a).capableof)) ∧
      raut (a, g, ps) ⇔ (∃ c : Agent • achieves (c, g, ps)) ∧
      (∃ p : ps • (extdes a).resourcesofplan p ⊆ (extdes a).ownedresources) ∧
      saut (a, g, ps) ⇔ aaut (a, g, ps) ∧ raut (a, g, ps) ∧
      achieves (a, g, ps) ⇔ g ∈ (extdes a).goals ∧
          (g, ps) ∈ (extdes a).planforgoal ∧ ps ≠ { }
```

In the definition of achieves, the expression $g \in (extdes\ a).goals$ states that an agent can only reason with respect to a set of plans associated with a *current* goal (i.e. one that it desires). However, in the original description, there is ambiguity about whether this must be so. The mathematical definitions make no mention of whether this proviso is part of the mechanism. If we are guided by the examples given by Sichman, however, it would appear that this proviso is, in fact, included.

Using the formal framework ensures that we are precise and unambiguous about any definitions presented within it. This is particularly important in this case, since whichever definition is used has ramifications for social dependence network categorisations. This is explored more fully in section 4.4.

According to these definitions, if agents are autonomous, then they may not *depend*, for resources or actions, on other agents. Consequently, the fact that a pocket calculator has the resources and the actions necessary for adding some numbers makes it autonomous. (By contrast, we have argued elsewhere that autonomy is not simply action or resource dependence, but involves the ability to make one's own choices, to generate goals [8].)

Nevertheless, these notions are useful to a motivated agent since in some motivational contexts, knowledge of the dependencies that exist between agents is important. They provide information as to when a goal can be satisfied (by performing the actions in a plan) without involving any other agents. Naturally, a planning agent may decide to pursue a plan that *does* involve others even if able to carry it out alone, for reasons of, for example, laziness, distribution of responsibility, efficiency, and so on.

4.3 Dependence Relations

Now we can consider the types of dependencies that exist between agents. An agent, A, *a-depends* on another agent, B, for a given goal, g, according to some set of plans of another to achieve g, if it has g as a goal, is not *a-autonomous* for g, and at least one action used in this plan is in B's capabilities. An agent, A, *r-depends* on another agent, B, for a given goal, g, according to some set of plans of another to achieve g, if it has g as a goal, is not *r-autonomous* for g, and at least one instantiation used in this plan is owned by B. An agent, A, *s-depends* on another agent, B, for a given goal, g, if it *r-depends* or *a-depends* on B.

The first predicate in the schema below states that given two agents, a and b, a goal, g, and a set of plans according to which a is not *a-autonomous* with respect to g, a *a-depends* on b for g with respect to ps, if and only if there is some agent, c, with the goal, g, and plans to achieve g, ps, such that at least one plan in ps has an action in the capabilities of agent b.

DependencyRelations

AutonomyRelations

adep _, rdep _, sdep _ : $\mathbb{P}(Agent \times Agent \times Goal \times \mathbb{P} \, Plan)$

$\forall a, b : Agent; g : Goal; ps : \mathbb{P} \, Plan \mid a \neq b \land (g \in (extdes \, a).goals) \bullet$
  adep $(a, b, g, ps) \Leftrightarrow \neg$ aaut $(a, g, ps) \land$
    $(\exists c : Agent \bullet$ achieves $(c, g, ps) \land \bigcup\{p : ps \bullet \text{ran } p\} \cap$
      $(extdes \, b).capableof \neq \{\}) \land$
  rdep $(a, b, g, ps) \Leftrightarrow \neg$ raut $(a, g, ps) \land$
    $(\exists c : Agent \bullet$ achieves $(c, g, ps) \land$
      $(\exists p : ps \bullet ((extdes \, c).resourcesofplan \, p) \cap$
        $(extdes \, b).ownedresources \neq \{\})) \land$
  sdep $(a, b, g, ps) \Leftrightarrow$ adep $(a, b, g, ps) \lor$ rdep (a, b, g, ps)

This reformulation also highlights some difficulties. As stated earlier, at no point is it made clear whether two agents can share an action or a resource. Second, it makes little sense to say that I *a-depend* on an agent for some goal if the actions that achieve that goal are in my capabilities. Similarly, it also makes little sense to say that I *r-depend* on some agent for some resource if that resource is also owned by myself. A more intuitive definition might be

adep $(a, b, g, ps) \Leftrightarrow (\exists\, c : Agent \bullet$ achieves $(c, g, ps) \land$
$(\exists\, x : \bigcup\{p : ps \bullet$ ran $p\} \bullet x \in (extdes\, b).capableof \land x \notin (extdes\, a).capableof))$

However, even when an agent is capable of some action of which I am not capable, and which I require for some plan, it again makes little sense to say there is a dependency. It is more appropriate to say that there is a possibility of that agent being able to help in achieving a goal. There is no doubt that such reasoning will be useful in certain situations. A better notion of actual *dependency* with respect to a goal, would be if *every* plan in the set of plans required some agent's assistance. In this respect there would be a real dependency on this agent in order to achieve the goal.

adep $(a, b, g, ps) \Leftrightarrow (\exists\, c : Agent \bullet$ achieves $(c, g, ps) \land$
$(\forall\, p : ps \bullet \exists\, x : $ ran $p \bullet x \in (extdes\, b).capableof \land x \notin (extdes\, a).capableof)))$

These relations provide an agent with the structures that can be used to reason about others with a view to choosing an appropriate course of action in the context of its dependencies on others' goals, plans, resources, and so on.

4.4 Dependence Situations

Sichman proceeds to use these relations to classify distinct *dependency relations* which arise. This subsection considers these situations. We must first note that there is an ambiguity between Sichman's mathematical and textual descriptions of dependency [12]. In the mathematical description, the dependency refers to any set of plans which any agent has, but the textual description refers only to the plans of the reasoning agent. In the interpretation that follows, we adopt the more restrictive version since it is consistent with the given notions of *independence* and *unilateral dependence* discussed later, and is more intuitive in reflecting the nature of autonomous agents. (We might equally have chosen the other alternative, however.)

Consider the situation where we have two agents, A and B, where A is not *a-autonomous* for some goal, g_1, according to A's plans, ps_1, to achieve g_1. We can then recognise the following situations.

A is *independent* with respect to B for g_1 if, according to ps_1, it infers that it does not *a-depend* on B for g_1.

A is *unilaterally dependent* on B if, according to ps_1, A *a-depends* on B, but there is no goal for which B *a-depends* on A.

Two agents are *mutually dependent* if they *a-depend* on each other for the same goal g_1 according to ps_1.

If, in addition, B is not *a-autonomous* for some goal, g_2, according to A's plans to achieve g_2 then we can also write the following.

Two agents are *reciprocally dependent* if they *a-depend* on each other for two different goals, g_1 and g_2, according to two sets of plans, ps_1 and ps_2 respectively.

Thus, given two agents, A and B, where A is not *a-autonomous* for some goal, g, we define the previous dependence situations with respect to g in the schema below.

DependencySituations

DependencyRelations
ind _, ud _ : $\mathbf{P}(Agent \times Agent \times Goal)$
md _ : $\mathbf{P}(Agent \times Agent \times Goal \times \mathbf{P}\ Plan)$
rd _ : $\mathbf{P}(Agent \times Agent \times Goal \times Goal \times \mathbf{P}\ Plan \times \mathbf{P}\ Plan)$

$\forall a, b : Agent;\ g_1, g_2 : Goal;\ ps_1, ps_2 : \mathbf{P}\ Plan \mid$
  $(a \neq b \wedge \mathsf{achieves}\,(a, g_1, ps_1) \wedge \neg\ \mathsf{aaut}\,(a, g_1, ps_1) \wedge g_1 \neq g_2)\ \bullet$
    $\mathsf{ind}\,(a, b, g_1) \Leftrightarrow \neg\ \mathsf{adep}\,(a, b, g_1, ps_1) \wedge$
    $\mathsf{ud}\,(a, b, g_1) \Leftrightarrow \mathsf{adep}\,(a, b, g_1, ps_1) \wedge$
       $\neg\,(\exists\,g : Goal;\ ps : \mathbf{P}\ Plan \mid \mathsf{achieves}\,(a, g, ps) \bullet \mathsf{adep}\,(b, a, g, ps)) \wedge$
    $\mathsf{md}\,(a, b, g_1, ps_1) \Leftrightarrow \mathsf{adep}\,(a, b, g_1, ps_1) \wedge \mathsf{adep}\,(b, a, g_1, ps_1) \wedge$
    $\mathsf{rd}\,(a, b, g_1, g_2, ps_1, ps_2) \Leftrightarrow \mathsf{achieves}\,(a, g_2, ps_2) \wedge$
         $\mathsf{adep}\,(a, b, g_1, ps_1) \wedge \mathsf{adep}\,(b, a, g_2, ps_2)$

These definitions would be more sensible if they were based on dependencies for actions which an agent does not have. For example, if A is independent of B, it implies that there is no way that B could *help* A in performing an action. A more intuitive definition of *independence* would be that A does not *need B*.

Consider the definition of mutual dependence between A and B. It states that A and B both have a goal g_1, and according to A's plans to achieve g_1, there is some plan in which B could perform an action, and some plan (not necessarily the same plan) in which A could perform an action. What this categorisation describes is a potential for cooperation. A more intuitive definition of mutual dependence would be that every plan in the set *needs* both agents.

Reciprocal dependence occurs when, according to two sets of plans, A could help B achieve some goal $g1$, and B could help A achieve some goal $g2$. However, since the definition is with respect to A's plans, if we assume that agents can only reason with respect to sets of plans associated with a desired goal (as suggested by the authors), it must be that A currently desires *both* goals g_1 and g_2. This is very restrictive since it rules out the possibility of bargaining when A has only one goal, for example, and B has only one other goal. In such a case, both agents may then help each other by adopting the other's goals.

The mechanism is described as *social exchange*, and the authors state that "one of them will have to adopt the other's goal first in order to achieve his own one in the future". As we have seen in one interpretation of this mechanism, A must necessarily have both goals, so this scenario is inappropriate. Even then, it is too restrictive since both plans may be carried out concurrently.

4.5 Local and Mutual Belief

The dependencies described above can be *locally* or *mutually* believed. A dependence is *local* if it only exists with respect to A's plans but not with respect to B's, and it is *mutual* if it occurs with respect to both A's and B's plans.

DependencySituationsLocalandMutual
_____

DependencySituations

lbmd _, mbmd _ : $\mathbb{P}(Agent \times Agent \times Goal)$

lbrd _, mbrd _ : $\mathbb{P}(Agent \times Agent \times Goal \times Goal)$

$\forall a, b : Agent; g_1, g_2 : Goal; ps_1, ps_2, ps_3, ps_4 : \mathbb{P} \ Plan \mid$

$\quad a \neq b \wedge g_1 \neq g_2 \wedge$ achieves $(a, g_1, ps_1) \wedge$ achieves $(a, g_2, ps_2) \wedge$

achieves $(b, g_1, ps_3) \wedge$ achieves $(b, g_2, ps_4) \wedge \neg$ aaut $(a, g_1, ps_1) \bullet$

$\quad$ lbmd $(a, b, g_1) \Leftrightarrow$ md $(a, b, g_1, ps_1) \wedge \neg$ md $(a, b, g_1, ps_3) \wedge$

$\quad$ mbmd $(a, b, g_1) \Leftrightarrow$ md $(a, b, g_1, ps_1) \wedge$ md $(a, b, g_1, ps_3) \wedge$

$\quad$ lbrd $(a, b, g_1, g_2) \Leftrightarrow$ rd $(a, b, g_1, g_2, ps_1, ps_2) \wedge \neg$ rd $(a, b, g_1, g_2, ps_3, ps_4) \wedge$

$\quad$ mbrd $(a, b, g_1, g_2) \Leftrightarrow$ rd $(a, b, g_1, g_2, ps_1, ps_2) \wedge$ rd $(a, b, g_1, g_2, ps_3, ps_4)$

More problems arise here, too. Notice, in particular, that both *local* and *mutual* belief require an analysis of both A's and B's plans, thus contradicting the following claim:

> "An agent locally believes a given dependence if he uses exclusively his *own plans* when reasoning about the others ..." [12]

In a subsequent paper, the authors drop the *hypothesis of external description compatibility* and instead concentrate on how they might detect *agency level inconsistency* resulting from two agents having different external description entries regarding each other [13]. However, it is also noticeable that their definition of mutually believed mutual dependence (MBMD) bears little relation to that proposed in the earlier paper [12]. The later definition is as follows:

> "As an example, if i infers a MBMD between himself and j for a certain goal g, this means he believes that (i) both of them have this goal and at least one plan to achieve it (ii) there is an action needed in this plan that he can perform and j can not perform (iii) there is an action needed in this plan that j can perform and he can not perform. " [13]

But the mathematical definition provides a different account of mutually believed mutual dependence: A and B both have goal g; according to A's plans A and B are not *a-autonomous* with respect to g; according to B's plans A and B are not *a-autonomous* with respect to g; there is some plan of A's which contains an action which B can do and a plan (possibly the same) which contains an action which A can do, and there is some plan of B's which contains an action which A can do and a plan (possibly the same) which contains an action which B can do.

In particular, the mathematical definition is *not* given in terms of an action not being in some agent's capabilities and, further, there is no mention of a *particular* plan within the set of plans, as required by the textual description above. This is precisely the kind of inconsistency we hope to avoid by specifying the mechanism formally within our framework, since we are then able to provide a unified and complete account of a system.

In the same paper [13], the authors do not assume the *hypothesis of external description compatibility* but state the following.

"For simplicity, let us consider that the plans of the agents are the same and both of them know the plans of the other."

Consequently, there can be agent level consistency in terms of what agents believe about the capabilities, resources and goals of each other, assuming they *know* the plans of every agent. This is evidently very useful, even though severely limiting, and further work explores agent reasoning about this class of problem.

5 A New Proposal

In this section we briefly describe a new proposal for external descriptions which allows for true autonomy (in the sense that an agent can never know the goals, actions and plans of another), simultaneous actions, active and non-active goals and plans, partial plans and a richer understanding of the social relationship between the entities in the world. In this respect, we can re-formulate the useful work of social dependence networks within a well-defined formal framework for agency and autonomy.

Consider fixing a screw into a block of wood. According to our hierarchical framework, this may require two agents: a screwdriver, and someone with the ability to use the screwdriver, both agents performing an action simultaneously. Every action in a plan must either be associated with the entity intended to to perform the action, or be associated with no entity if the entity involved in its instantiation has not yet been chosen.

In the following example, we illustrate our new representation of a plan for use in external descriptions. It consists of an action which I will perform, followed by two actions performed by two entities simultaneously, followed by three actions performed simultaneously by three entities (including one by me), followed by some action to be performed by an as yet unknown entity.

$$\langle \{(a_1, \{me\})\}, \ \{(a_{2_1}, \{entity1\}), (a_{2_2}, \{entity2\})\},$$
$$\{(a_{3_1}, \{me\}), (a_{3_2}, \{entity2\}), (a_{3_3}, \{entity4\})\}, \ \{(a_4, \{\})\}\rangle$$

A plan, therefore, has the following new type.

$NewPlan == \mathsf{seq}\,(\mathbf{P}(Action \times (\mathbf{P}\ Entity)))$

The schema below specifies an external description which includes the *current* goals of an agent. Associated with each such goal are a set of plans which together form the set of current plans. In addition, we also define the set of *all* goals of an agent — some of which are currently desired and some of which are not — and the corresponding set of *all* plans. (Note that in this respect we can be clear about categorisations based on certain types of plans, goals and so on.)

In addition, we define three useful but redundant variables which for each plan return the set of action-entity pairs, actions, and entities involved in the plan, respectively. The last predicate ensures that given an action-entity pair in a plan where the entity is defined, the action must be in the capabilities of the entity.

NewExternalDescription
Agent
plans : $\mathbb{P}$ NewPlan
allgoals : $\mathbb{P}$ Goal
allplans : $\mathbb{P}$ NewPlan
planforgoal : Goal $\rightarrowtail$ $\mathbb{P}$ NewPlan
actionsofplan : NewPlan $\longrightarrow$ $\mathbb{P}$ Action
entitiesofplan : NewPlan $\longrightarrow$ $\mathbb{P}$ Entity
actionentities : NewPlan $\longrightarrow$ $\mathbb{P}$(Action × $\mathbb{P}$ Entity)

goals $\subseteq$ dom planforgoal
plans = $\{p : NewPlan; g : Goal \mid g \in goals \land p \in planforgoal\ g \bullet p\}$
allgoals = dom planforgoal
allplans = $\bigcup$(ran planforgoal)
$\forall\ p : NewPlan \bullet actionsofplan\ p = \{aes : actionentities\ p \bullet first\ aes\} \land$
$\quad entitiesofplan\ p = \bigcup\{aes : actionentities\ p \bullet second\ aes\} \land$
$\quad actionentities\ p = \bigcup(ran\ p) \land (\forall\ aes : actionentities\ p; e : Entity \mid$
$\quad\quad second\ aes = \{e\} \bullet first\ aes \in e.capableof)$

Further work can then progress using the definitions given above to provide new social dependence network categorisations based on the original formalisms. As a small example, we can say that an agent is *t-autonomous* with respect to a plan if all the actions the plan contains are within its own capabilities. Essentially:

taut $(agent, plan) \Leftrightarrow actionsofplan\ plan \subseteq (agent.capableof)$

6 Conclusions

Social Dependence Networks are a valuable source of information about the relationships within a multi-agent world, and provide the necessary structure that can be exploited by agents in order to function effectively. They allow agents to reason about resources and capabilities of others in order that they may enter into a negotiation to persuade these others to assist them in completing their tasks. This paper has described the work of Sichman et al. in developing computational models of dependence networks and has reformulated it in another, formal, framework. By reformulating it in these terms, we have been able to identify deficiencies in the original characterisation of dependence networks and the external descriptions of agents within the networks. We have addressed these deficiencies, and offer a modified view of external descriptions which removes much of the ambiguity and presents a stronger and more consistent, formal model which can easily be extended to define social dependence networks.

In reformulating dependence networks in this way, we have also presented a case study which shows how the formal framework that we have previously developed can be applied to provide an environment in which we can describe and reason about theories and models of MAS. Moreover, we have highlighted inconsistencies and ambiguities, and outlined how such models may be incorporated within our framework.

Acknowledgements Many thanks to Rafael Bordini for detailed comments and suggestions on an earlier version of this paper.

References

1. C. Castelfranchi. Social power. In Y. Demazeau and J. P. Muller, editors, *Decentralized Artificial Intelligence*, pages 49–62. Elsevier North Holland, 1990.
2. C. Castelfranchi, M. Miceli, and A. Cesta. Dependence relations among autonomous agents. In E. Werner and Y. Demazeau, editors, *Decentralized Artificial Intelligence*, pages 215–231. Elsevier North Holland, 1992.
3. I. Craig. *Formal Specification of Advanced AI Architectures*. Ellis Horwood, 1991.
4. M d'Inverno and J. Crowcroft. Design, specification and implementation of an interactive conferencing system. In *Proceedings of IEEE Infocom, Miami, USA. Published IEEE*, 1991.
5. M. d'Inverno and M. Priestley. Structuring a Z specification to provide a unifying framework for hypertext systems. In J. P. Bowen and M. G. Hinchey, editors, *ZUM'95: 9th International Conference of Z Users, Lecture Notes in Computer Science*, pages 83–102, Heidelberg, 1995. Springer-Verlag.
6. R. Goodwin. Formalizing properties of agents. Technical Report CMU-CS-93-159, Carnegie-Mellon University, 1993.
7. M. Luck and M. d'Inverno. Engagement and cooperation in motivated agent modelling. In *Proceedings of the first Australian DAI Workshop*. Springer Verlag, 1995.
8. M. Luck and M. d'Inverno. A formal framework for agency and autonomy. In *Proceedings of the First International Conference on Multi-Agent Systems*, pages 254–260. AAAI Press / MIT Press, 1995.
9. M. Luck and M. d'Inverno. Goal generation and adoption in hierarchical agent models. In *AI95: Proceedings of the Eighth Australian Joint Conference on Artificial Intelligence*. World Scientific, 1995.
10. M. Luck and M. d'Inverno. Structuring a Z specification to provide a formal framework for autonomous agent systems. In J. P. Bowen and M. G. Hinchey, editors, *ZUM'95: 9th International Conference of Z Users, Lecture Notes in Computer Science*, pages 48–62. Springer-Verlag, 1995.
11. B. G. Milnes. A specification of the Soar architecture in Z. Technical Report CMU-CS-92-169, School of Computer Science, Carnegie Mellon University, 1992.
12. J. S. Sichman, Y. Demazeau, R. Conte, and C. Castelfranchi. A social reasoning mechanism based on dependence networks. In *ECAI 94. 11th European Conference on Artificial Intelligence*, pages 188–192. John Wiley and Sons, 1994.
13. J. S. Sichman and Yves Demazeau. Exploiting social reasoning to deal with agency level inconsistency. In *Proceedings of the First International Conference on Multi-Agent Systems*, pages 352–359. AAAI Press / MIT Press, 1995.
14. J. M. Spivey. *The Z Notation*. Prentice Hall, Hemel Hempstead, 2nd edition, 1992.
15. M. J. Wooldridge and N. R. Jennings. Applying agent technology. *Journal of Applied Artificial Intelligence, special issue on Intelligent Agents and Multi-Agent Systems*, To appear, 1995.

A Introduction to Z

The formal specification language Z is based on set theory, first order logic and predicate calculus. It extends the use of these languages by allowing an additional mathematical type known as the *schema type*. Z schemas have two parts: the upper, declarative, part which declares variables and their types, and the lower, predicate, part which relates and constrains those variables. The type of any schema can be considered as the cartesian product of the types of each of its variables, without any notion of order, but constrained by predicates. Modularity is facilitated in Z by allowing schemas to be included within other schemas. We can select a state variable, *var*, of a schema, *schema*, by writing *schema.var*.

To introduce a type in Z, where we wish to abstract away from the actual content of elements of the type, we use the notion of a *given set*. We may write *NODE* to represent the set of all nodes. If we wish to state that a variable takes on some set of values or an ordered pair of values we write $x : \mathbf{P}\,NODE$; $x : NODE \times NODE$, respectively. The generic functions *first* and *second* return the first element and second element of any ordered pair, respectively.

A *relation* type expresses some relationship between two existing types, known as the *source* and *target* type. When no element from the source type can be related to two or more elements from the target type, the relation is a *function*. A *total* function ($\longrightarrow$) is one where every element in the source set is related, while a *partial* function ($\nrightarrow$) is where not every element in the source is related. Finally, A sequence (seq)is a special type of function where the domain is the contiguous set of numbers from 1 up to the number of elements in the sequence. For example, consider the following function which defines a relation between nodes: $Rel = \{(node1, node2), (node2, node3), (node3, node2), (node4, node4)\}$.

The *domain* (dom) of a relation or function is those elements in the source set which are related, and the *range* (ran) is those elements in the target set which are related. In this case dom $Rel = \{node1, node2, node3, node4\}$ and ran $Rel = \{node2, node3, node4\}$.

Sets of elements can be defined using set comprehension. For example, the following expression denotes the set of squares of natural numbers greater than 10 $\{x : \mathbb{N} \mid x > 10 \bullet x * x\}$. The way to write down predicates in Z is nonstandard. To state that, say, any number greater than 10 has a square greater than 100, we write: $\forall n : \mathbb{N} \mid n > 10 \bullet n * n > 100$.

Lastly, we make use of *mapseq*, which takes a function and a sequence and applies the function to each element of the sequence and *mapset*, which takes a function and a set and applies the function to each element of the set.

$$
\begin{array}{l}
\underline{\quad[X, Y]\quad} \\
\quad mapseq : (X \nrightarrow Y) \nrightarrow \text{seq}\,X \nrightarrow \text{seq}\,Y \\
\quad mapset : (X \nrightarrow Y) \nrightarrow \mathbf{P}\,X \nrightarrow \mathbf{P}\,Y \\
\hline
\quad \forall\, seqs : \text{seq}\,X;\ xs : \mathbf{P}\,X;\ fun : X \nrightarrow Y \bullet \\
\qquad mapseq\ fun\ seqs = \{n : \mathbb{N} \mid n \in 1\,..\,\#seqs \bullet (n, fun\,(seqs\,n))\} \wedge \\
\qquad mapset\ fun\ xs = \{x : X \mid x \in xs \bullet fun\,x\}
\end{array}
$$

From Practice to Theory in Designing Autonomous Agents

Lorenzo Sommaruga, Nadia Catenazzi

Intelligent Agents Lab. (LAI), Computer Science Dept.,
Universidad Carlos III de Madrid, Spain
Butarque 15, 28911 Leganés MADRID
tel. +34-1-624 9416, fax: +34-1-624 9430
e-mail: {los, nadia}@ing.uc3m.es

Abstract

The purpose of this paper is to introduce some principles, in the form of assumptions, which could be useful to consider in developing agents architecture. These principles provide design guidelines concerning agent and group structuring, communication, decision making, and computational aspects for agent cooperation. They generate requirements and constraints for cooperative agents development.

These principles have been derived from the CooperA experience in designing and implementing architectural frameworks for agent cooperation. In particular, basic characteristics of autonomous agents have been formalised in a computational model.

Keywords: Agent architecture, cooperation, theoretical foundations, multi-agent interaction.

1 Introduction

The objective of this paper is to describe our experience from the practice to the theory of agents architectures design. In general, the developer of a system for the cooperation of a group of agents has to take into account a number of issues concerning task allocation, communication, decision making, knowledge structures selection, coordination/cooperation, etc.

The principles adopted represent a general guidance to the development phase of such architecture, from which requirements and constraints for cooperative agents emerge. They have been derived from our practical experiences in the field. We have designed and implemented two architectural frameworks for agent cooperation: CooperA [Avouris-et-al89][Sommaruga-et-al95], a multi-agent system with rich inter-agent communication and heterogeneous knowledge representation features; and CooperA-II [Sommaruga&94], an evolution of the first CooperA architecture to a workbench for the cooperation of autonomous agents based on domain independent cooperative heuristics. Both approaches are based on the reciprocal awareness of activities and requirements between agents, referred to as skills and needs. CooperA-II is the result of a 5-year research effort [Sommaruga-et-al95], (carried out at CEC JRC - Ispra, Italy, and the Psychology Department (AI group) of the University of Nottingham), which demonstrates an increased architectural complexity and sophistication, and supports the validity of the principles proposed in this paper.

2 The Architecture Design

In the design of an architecture for cooperative autonomous agents we have introduced a stratified view for a cooperative agent model, in order to help a programmer identify and organise distinct levels of functionalities within an agent. The concept of competences is very important in this context. It deals with the abstraction of problem tasks and subtasks in terms of the skills and needs of an agent. A full description of the CooperA-II's methodological approach is reported in [Sommaruga93] together with the research results.

We distinguished four levels:

Real World level: It is the lowest level which deals with the real world, e.g. a real-life problem the agent is supposed to interact with;

Application level: It concerns the computational (expert) systems, e.g. computer programs which model the problem;

Abstraction level: It represents the competence abstraction, such as the definition of the agents' models through their skills and needs (cf. also [Sommaruga93]);

Cooperation level: It identifies the level of cooperation, where the agent control resides.

Fig. 1. Levels and steps for defining an architecture of cooperative agents.

In the Application level we find the application-dependent part of the system/agent (application system). For example, a procedural code module, or other software entity like an expert system may be considered as an application part of the agent. This part of a system is considered the basic entity which has to participate and interact with other entities in reaching the solution of a problem. The application system of an agent deals with the domain capabilities the agent can offer, the effects it can generate on its external environment (real world and other agents) and the requirements and inputs it needs. We usually refer to these as goals and requirements or skills and needs of the agent. It is clear, from this description of the Application level, that it has the full richness of the KADS model [Wielinga-et-al92].

It is interesting to note that up to the Application level, there are no real innovations and differences with respect to a traditional computational (expert) system. Up to the Abstraction level, frameworks already existed, like MACE [Gasser-et-al87] and CooperA. They have already considered, in different forms, the abstraction of the competences of an agent.

In the Abstraction level the step of creating abstractions can be seen similarly to Newell's attempt to specify the knowledge level for agents and the use of representation [Newell92] (see also [Aitken-et-al94]). A representation is a structure which realises the knowledge at a symbolic level. "The nature of the approximation is such that the representation at the symbol level can be seen as knowledge plus the access structure to that knowledge" [Newell92].

In the CooperA-II model, the competence description covers the abstraction of a problem in terms of tasks, subtasks, and their relations which determines how the problem can be solved, their distribution to different agents and the run time interactions between the agents. The description of a problem is determinant on its decomposition, its distribution and hence how it behaves when solving a problem cooperatively. The decomposition process of a problem happens firstly at the application level during its definition, and secondly it is applied during the problem solving phase for allocating activity to the agents. This assignment ranges from being completely determined a-priori to being opportunistic and dynamically assigned. Non-adaptive versus adaptive organisations emerge according to the type of assignment.

Abstraction and cooperation are certainly the two most important levels of our architecture. In fact, they allow the transformation of a simple (expert) system into a social agent open to collaborating with similar ones.

The Cooperation level of an agent is actually considered as a meta level because it works mainly on meta information such as the abstraction of competences or other knowledge structures which represent, for instance, knowledge about knowledge or about other agents.

2.1 The Agent Model

The basic concept of the architecture is the agent. First of all, we provide a definition of sets which are used in the formalisation of an agent, following a system theory approach [Zadeh&69].

The agent model consists of two elements: the agent itself and its environment. The model of an agent is graphically presented in Fig. 2.

In the environment, we distinguish two general sets, common for a whole group of agents, and independent from the agent: R and C.

R is the real world, which can receive outputs from the agents (e.g. the execution of an agent action), and can provide inputs to the agents (e.g. data/observations of the real world).

C is a set representing the communication channel, established between all agents. It is composed of messages. It is assumed that the communication channel uses a message passing mechanism. This is compared to a shared memory technique which is considered inadequate for a real spatial distribution of agents [Sommaruga93].

Concerning the agent itself, its inputs and outputs are collected respectively in the sets I and O. Both of them consider inputs from, or outputs to, two different locations: R and C. An input can come from the real world through sensors, or it can also come from the communication channel C in the form of a message from another agent. Similarly, an output can be directed to the real world in the form of an action of an actuator or to C in the form of a message for another agent.

Other sets refer to internal knowledge which is local to each agent. However, they introduce general concepts adopted in our model. The set S is the set of skills of the agent. It contains a description of information or tasks which can be provided by the agent. The set N is the set of needs of an agent, which enumerates any information the agent wants to receive. This information represents what an agent needs in order to satisfy its interests. The sets S and N constitute an abstraction of the competences of the agent. These sets may be restricted to static sets fixed at the definition time of the agent.

Agent **Environment**

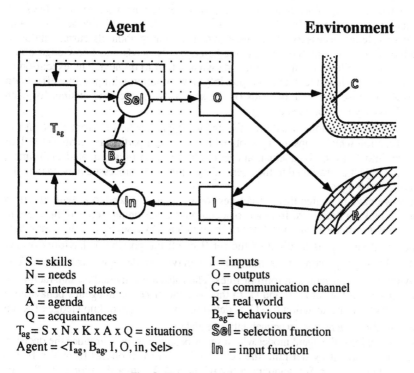

S = skills
N = needs
K = internal states
A = agenda
Q = acquaintances
T_{ag} = S x N x K x A x Q = situations
Agent = <T_{ag}, B_{ag}, I, O, in, Sel>

I = inputs
O = outputs
C = communication channel
R = real world
B_{ag} = behaviours
Sel = selection function
In = input function

Fig. 2. The CooperA-II Agent Model.

A number of sets concern dynamic knowledge of the agent, which can vary during the interaction with others and may influence cooperation of the agent. They include the set K of internal states, the agenda set A and the acquaintances set Q.

The set K is composed of the internal states of an agent. A state of an agent may be composed, for example, of the current status of knowledge of the agent skills and needs, of messages managed by the agent and other attributes.

The set A contains all the current potential actions of an agent, collected in an agenda structure. The name agenda underlines the nature of this set which is derived from the Latin word for "(things) to be done." In CooperA-II, the actions are described in the form of acts. These acts relate to computing information locally or executing a task, supplying information to other agents, being informed, and requesting information from another agent. They are functional derivatives of the definitions of the Inform, Request, and Cause-to-Want operators introduced by the Speech Acts theory [Cohen&79]. They are fully described in [Sommaruga93] and [Sommaruga-et-al95], together with a case-based-like semantics of agent acts.

The set Q contains all the acquaintances of the agent in form of structured knowledge elements about other agents. It consists of knowledge about the other agents in the group, such as information about their skills and their needs.

The set S of skills, the set N of needs, the set K of states of an agent, the set A of agenda and the set Q of acquaintances compose together what we refer to as the set T of situations of the agent. In other words $S \times N \times K \times A \times Q = T$. The set T contains all the knowledge necessary for an agent when reasoning about its current status of knowledge and its potential actions, mediated by its needs and knowledge about others. Thus, it allows the agent to decide what action to do next. The execution of such actions will lead the agent to a possibly new situation with possible effects on C and R. Because the set T depends on the sets S, N, Q, etc. it is distinct for each agent. This may be expressed as T_{ag}.

Under these definitions, we define the agent's input function *In* from T x I into T. *In* is a function which maps an input object $i \in I$ under a particular situation $t \in T$ of the agent into a new derived situation $t' \in T$. The function *In* is activated each time a new input object is presented to the agent. This input when processed may change the situation of the agent.

Once we have modelled these structures for an agent, we need to provide it with a means of making decisions. In order to accomplish a reasoning step and therefore to decide how to act, an agent model has to be completed by a set B_{ag} of the behaviours of the agent ag. B_{ag} is a subset of the set B of all the possible behaviours of any agent. The set B_{ag} contains a number of behaviours the agent can adopt. The behaviours are represented in form of rules. The rules are activated on satisfaction of conditions about the set K, A and Q. In this way the effect of an agent's behaviour b $\in B_{ag}$ which can be an action $o \in O$, is determined by a status of knowledge of K, an agenda A of the agent, and an acquaintance model Q.

Once completed the agent model with a set of behaviours, we introduce the decision making of an agent by defining the action selection function *Sel* from T x B into $T\square x \square O$. The function *Sel*, applying a behaviour $b \in B_{ag}$ to a situation $t_0 \in T$ of an agent, produces a possible new situation $t_1 \in T$ and an output $o \in O$, which can be an action $r \in R$ or a message $c \in C$ to other agents. *Sel* is a function which is activated each time a new situation is reached by the agent. This new situation can then evolve, possibly producing an output action. In CooperA-II, the *Sel* function, implemented as a forward-reasoning mechanism in the controller of each agent, allows an agent to coordinate and plan tasks using the heuristic knowledge about cooperation, expressed in the behaviours B_{ag}. This knowledge has been designated as a Cooperative Heuristics Knowledge Base (CH-KB). The CH-KB contains knowledge at the meta level of control which generates agent decisions relating to *how* and *when* to act, and *what* goal to achieve. This knowledge expresses a number of different possible cooperative behaviours for an agent, which include, for example decomposing and distributing tasks, and collaborating spontaneously or on request.

In summary, on the basis of the above definitions, an agent is a tuple:
$\langle T_{ag}, B_{ag}, I, O, In, Sel \rangle$.

2.2 The Functional Behaviour of an Agent

The functional behaviour of a system generally relates to how the system evolves, describing how it works.

In order to formalise the functional behaviour of an agent we consider an initial state t_0 in which an agent ag = $\langle T_{ag}, B_{ag}, I, O, In, Sel \rangle$, with a behaviour b receives an input i_0.

The function *In* is applied on the input i_0 received by the agent in the situation t_0. It transforms t_0 into t_0'. The new situation t_0' activates the function *Sel* also on b. This provides the transformation of t_0' into a new situation t_1 and a possible output o_1. This output modifies the real world R or the communication channel C. Thus, it may generate a new input from R through the execution of an action, or from C by another similar agent. The cycle can then repeat.

In other words, we have:

$$i_0 \in I, t_0, t_0', t_1 \in T_{ag}, b \in B_{ag}, o_1 \in O$$
$$In(i_0, t_0) = t_0'$$
$$Sel(t_0', b) = \langle t_1, o_1 \rangle$$
$$Sel(In(i_0, t_0), b) = \langle t_1, o_1 \rangle$$

We observe that the input received may leave the current situation unchanged. In this case, the function *In* will return the same situation given in input ($In(i_0, t_0) = t_0$), and the *Sel* function will return the same situation received in input and a null action, i.e. no action, ($Sel(t_0, b) = \langle t_0, - \rangle$).

In order to generalise this process, we observe that the function *Sel* from T x B to T x O, where $B_{ag} \subseteq B$, can map any internal situation of any agent and its behaviour to a new internal situation with a possible output action. This is formalised as:

$$i_n \in I, t_n, t_{n+1} \in T_{ag}, b \in B_{ag}, o_{n+1} \in O$$
$$Sel(In(i_n, t_n), b) = \langle t_{n+1}, o_{n+1} \rangle.$$

We showed above that the firing of the heuristics ($b \in B_{ag}$) is based on the current situation t_0' of an agent: $Sel(t_0', b) = \langle t_1, o_1 \rangle$. The situation includes a set of possible acts of an agent collected in the agenda.

2.3 The Cooperative Architecture Model

Once a model of the basic components for an architecture of agents is defined, the aggregation of such parts is immediate and constructive.

The composition of such agents within an environment generates a cooperative architecture. Fig. 3 presents a cooperative architecture restricted to the case of two agents.

The common sets C and R of the environment, external to the agent, allow us to integrate an agent object with other similar objects. This is a fundamental step for the creation of a distributed agent system.

All the agents in a group will be connected via the communication channel and eventually may act on the real world.

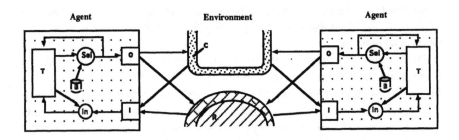

Fig. 3. The Cooperative Architecture Model.

On the basis of the previous definitions about the agent model, a cooperative architecture of agents is defined as: CA = <C, R, Ag>, where Ag is a set of agents defined according to the previous model.

This abstraction has been introduced in order to define a formal reference model for our principles. In section 3 we propose a formalisation of these principles (in the form of assumptions), for the definition of a cooperative agent architecture.

2.4 Implementation

The CooperA-II workbench system for the cooperation of autonomous agents has been implemented (about 5000 lines of Lisp code, excluding the interface) reflecting the model above described.

Two application examples in different real world domains were developed [Sommaruga93]. The Libra application consists of a group of four cooperating agents which simulate a librarian helping a user to find references about subjects, authors, etc. in a knowledge base of references. The Distributed Chemical Emergences Manager 2 (DChEM2) consists of a group of three agents which interact with a user in order to determine the nature of a substance involved in a possible chemical accident.

3 Principles for Cooperative Agents Design

Some of the principles are dictated by very general constraints on computational (intelligent) systems. For example, in [Newell82] Newell's Principle of Rationality already states that: "If an agent has knowledge that one of its actions will lead to one of its goals, then the agent will select that action." In particular, this principle is reflected in principle P.2. Others are influenced by more specific DAI and cooperative requirements and refer specifically to our CA.

P.0 - Agent Structure and Cooperative Architecture

An agent is a tuple $<T_{ag}, B_{ag}, I, O, In, Sel>$, where
$T_{ag} = S \times N \times K \times A \times Q =$ situations

S = skills	B_{ag} = behaviours
N = needs	I = inputs
K = states	O = outputs
A = agenda	In: T x I -> T is the input function
Q = acquaintances	Sel: T x B -> T x O is the selection function.

A Cooperative Architecture for agents is a tuple $<C, R, Ag>$, where
C = communication channel
R = real world
Ag = agents

P.1 - Communality Requirement

In order to have cooperation the agents must have overlap of skills and needs, i.e. for each agent the intersection between its set of skills and the set of needs of any other agent in Ag must not be empty. Assuming that S_{ag_i} is the set of skills of the agent ag_i, and N_{ag_j} is the set of needs of the agent ag_j:

$$\forall ag_i \in Ag, \exists ag_j \in Ag, ag_i \neq ag_j$$
$$S_{ag_i} \cap N_{ag_j} \neq \emptyset$$

P.2 - Action Selection

For an agent in a situation, the selection of the next best action to carry out depends on the particular behaviour defined for the agent. This is expressed in the function Sel from T x B to T x O which maps an internal agent's situation $In(i_n, t_n) \in T$ and the behaviour $b \in B$ of the agent to a new internal situation $t_{n+1} \in T$ with a possible output action $o_{n+1} \in O$, where $In(i_n, t_n)$ is a new situation returned by the function In applied to a situation $t_n \in T$ and an input $i_n \in I$ at instant n:

$$i_n \in I, b \in B_{ag}, t_n, t_{n+1} \in T_{ag}, o_{n+1} \in O$$
$$Sel (In(i_n, t_n), b) = <t_{n+1}, o_{n+1}>$$

P.3 - Communication Requirement

In a real distributed system the communication is useful for cooperation. Input and output messages can be managed asynchronously within an agent and with respect to the control of the agent. A common message structure is supposed for all the agent.

A protocol of communication can provide an act based semantics to the agents.

The use of a dictionary mechanism can allow distinct agents to maintain a conceptual (semantic) correspondence between a referent information (skill or need) at the application level and a referent to the same information at the cooperation level.

P.4 - Group Dynamics

The sets of capabilities, S (skills), and of necessities, N (needs) of a group of agents can be varied by changing the composition of the group of agents, or the set of skills or needs of a single agent.

The set S_{ag_i} of skills and the set N_{ag_i} of needs of the agent ag_i can defined and changed at run time for each agent, determining specific sets T_{ag_i}. The skills and the needs of the whole group of agents are given respectively by the composition of the set S_{ag_i} and the set N_{ag_i} of each agent. Therefore, by changing the composition of a group Ag of agents or the single skills/needs it is possible to change the sets of the skills and the needs of the whole group. Thus:

$$S = \{\forall ag_i \in Ag: \cup S_{ag_i}\}$$
$$N = \{\forall ag_i \in Ag: \cup N_{ag_i}\}.$$

P.5 - Self Initialisation

Groups of agents are open. Introducing a new agent in a group involves firstly updating the acquaintances. A flexible acquaintances model has to be dynamically created and maintained. Any agent must be able to introduce itself into a group, importing the skills and needs of others and exporting its own skills and needs to the group.

The set Q_{ag_j} of the acquaintances of the agent $ag_j \in Ag$ is a function of the sets of skills and needs of all the agents in the group. In other words:

$$Q_{ag_j} = F(S_{ag_i}, N_{ag_i}), \forall ag_i \in Ag$$

where F is a function which creates and updates the acquaintances according to changes in the group.

P.6 - Principle of Patience

The agent must be patient, i.e. must have the ability to wait for some change, from its internal world represented by T (e.g. changes in the agenda, set A) or from its external world through the inputs I (e.g. messages from other agents, C, or changes at the application level, R). Agents always pay attention to what they are told and embody a control régime of the form:

1: if *Change* (I, T) = True then goto 2
   else goto 1;
2: *Apply* (Sel (t, b))

where *Change* returns True if I or T changes, goto directs the interpreter to step 1 or 2, and *Apply* executes the selection function.

P.7 - Goal Induction

A common protocol for an act based semantics is used by all the agents in a group (see P.3). All the possible actions of an agent can be considered in the form of acts in an agenda. The semantics of these acts can provide control over other individuals, and therefore goal induction. In fact, under a particular behaviour $b \in B$, an agent's action, i.e. an output $o_{n+1} \in O$, in a situation t_n is determined according to P.2 by:

$$i_n \in I, b \in B_{ag}, t_n, t_{n+1} \in T_{ag}, o_{n+1} \in O$$
$$Sel(In(i_n, t_n), b) = <t_{n+1}, o_{n+1}>$$

It is evident that the agent's action o_{n+1} is a function of $i_n \in I$, where i_n can be a message from another agent which requests the recipient to accomplish a goal.

P.8 - Behaviours and Cooperative Heuristics

The particular behaviour of an agent, expressed in P.2, can be explicitly defined in the form of rules which embody knowledge about cooperation. These rules are called Cooperative Heuristics (CHs), and may be collected in a knowledge base (CH-KB) corresponding to B_{ag}.

The global effect which a sub set of Cooperative Heuristics produces in an agent is called the agent *micro behaviour* (m-behaviour). Various micro behaviours can be composed generating a *macro behaviour* of an agent, usually considered as the behaviour of the agent.

P.9 - Compositionality of Local m-behaviour

The overall behaviour of the agent ag is determined by the composition of the micro behaviours (m-behaviours) $b_i \in B_{ag}$ of the agent, where $B_{ag} \subseteq B$ is the set of all the m-behaviours of agent ag. The overall behaviour of an agent can be considered in terms of the possible effects generated by its micro behaviours. In other words, by assuming that:

$$E_{m\text{-}b_i} = \{\forall t_n \in T_{ag}, b_i \in B_{ag}: Sel(t_n, b_i) = <t_{n+1}, o_{n+1}>\}$$

is the set of all the possible effects generated by a micro behaviour $b_i \in B_{ag}$,

then we can assert that:

the set of all the effects generated by the overall behaviour B_{ag} of the agent ag is determined by:

$$E_{ag} = \{\ \forall b_i \in B_{ag}: \cup E_{m\text{-}b_i}\} =$$
$$= \{\forall t_n \in T_{ag}\ \forall b_i \in B_{ag}: Sel(t_n, b_i) = <t_{n+1}, o_{n+1}>\}$$

Thus, E_{ag} is the set of all the possible effects generated by all the micro behaviours $b_i \in B_{ag}$

P.10 - Compositionality of Global Behaviour

The global behaviour of a group of agents is determined by the composition of the behaviours of each agent in the group. In terms of the possible effects generated by the overall behaviour of each agent, and following P.5 we can state that:

the set of all the possible effects generated by all the behaviours of the agents in the group, i.e. the global effects set, is:

$$E_{Ag} = \{\forall ag_i \in Ag: \cup E_{ag_i}\} =$$
$$= \{\forall ag_i \in Ag\ \forall t_n \in T_{ag_i}\ \forall b_j \in B_{ag_i}: Sel(t_n, b_j) = <t_{n+1}, o_{n+1}>\}.$$

3.1 Remarks on the Principles

Having defined the principles of our Cooperative Agents Architecture, it is interesting to make some observations about their possible implications in the agent design.

The first principle (P.0) derive from the necessity to base all the other principles on a unique and well defined reference model of an agent.

P.1 is an essential principle for cooperation, because it guarantees the need for interaction of the agents (inter-activity).

P.2 is another basic principle which guarantees the local activity of an agent, by providing a mechanism for selecting the next best action (activity). For example, the adoption of this kind of selection algorithm leads to a design choice of a particular control cycle of an agent, which can guarantee a behaviour based control.

P.3 establishes a communication channel between agents. Asynchronous management of messages is considered in order to assure independence of the mere communication task from the control/decision-making one. Input and output messages or actions are managed by independent processes. For instance, an input queue manager handles all the incoming messages, while a different process, an output queue manager, autonomously dispatches the output messages.

In P.4 and P.5 we leave open the composition of an agents group providing a flexible, fault tolerant, and independent architecture. We observe that these principles have been extended in order to cope with agent skill learning.

The autonomous nature is also reflected by P.6 and P.7 where an agent, on one hand, is always alive and waiting for some change in its world, and, on the other hand, can filter the incoming requests. In fact, P.7 expresses not only that an agent can be reactive to changes, i.e. doing an action under an input (goal induction), but also, that its decision about the action depends on its own behaviour. Agents are not to be considered strictly benevolent just because they pay attention to their inputs; they also have a discerning capability for their actions.

The act triggering mechanism of the agenda could be used for managing, by means of synchronisation, continuous or non-instantaneous actions which are not explicitly dealt with in the model.

The agent behaviour is considered in the last three principles, which concern a specific cooperative architecture (CooperA-II), but may be generalised to other different approaches to control behaviour of agents. Our experience gives support to an explicit representation of the agent behaviour (P.8).

It is worth noting that while, locally to an agent, the m-behaviours influence the selection of the next action to carry out, as emerges from P.2 and P.9, globally, the behaviours (see P.8) and their coordination determine the integration of agents into a group and their interaction, as emerges from P.10.

4 Conclusions

The key contribution of this paper is the presentation of a theoretical model for cooperative agents development, which provides general requirements and constraints. The principles give design guidelines concerning agent and group structuring (P.0,□P.4, P.5), communication (P.3, P.0), decision making (P.2, P.7), agent control (P.6, P.7, P.2), coordination/cooperation (P.1, P.3, P.8), and behaviours abstraction (P.8, P.9, P.10).

In particular, with respect to the last point, thanks to the explicit representation of the agent behaviour (B_{ag}) in the CHs knowledge base, this knowledge could be easily extended or changed, modifying at the same time the behaviour of the agent.

Although the principles derive from a particular experience (the CooperA project), they can be generalised and used as general guidelines for designing and implementing an architecture for cooperative interaction of agents. The resulting architecture will be modular and flexible, because it maintains a distinction between the cooperation/control level from the application level, according to our layered view. Moreover, the underneath agent model results to be sufficiently general to be adopted in various domains and applications. We have successfully applied this architectural approach to Robotics, in particular to the definition of a workbench for the development and simulation of cooperative autonomous robots [Sommaruga-et-al95b]. Finally, we are applying this model to real mobile robots such as the Kephera [Mondada-et-al93] and Rug Warriors [Jones&93].

Acknowledgements

The authors wish to thank Prof. Nigel Shadbolt of the Psychology Dept., Nottingham University, for his contributions to the CooperA-II work.

References

[Aitken-et-al94]   J. S. Aitken, F. Schmalhofer, N.R. Shadbolt, "A Knowlegdge Level Characterisation of Multi-Agent Systems," in The 1994 Workshop on Agent Theories, Architectures, and Languages, ECAI-94, Amsterdam, August 1994.

[Avouris-et-al89]   N. Avouris, M. Van Liedekerke, L. Sommaruga: "Evaluating the CooperA Experiment: the transition from an Expert System Module to a D.A.I. Testbed for Cooperating Experts", presented at the 9th D.A.I. Workshop, Washington, USA, Sept. 1989.

[Cohen&79]   P.R. Cohen and C.R. Perrault, "Elements of a Plan-Based Theory of Speech Acts," Cognitive Science 3, 177-212, 1979.

[Gasser-et-al87]   L. Gasser, C. Braganza, N. Herman; "MACE: a flexible testbed for Distributed A.I. research," in "Distributed Artificial Intelligence," M. N. Huhns (editor), 119-152, Pitman Publishing, London, Morgan Kaufmann Publishers, San Mateo CA, 1987.

[Jones&93]   J. L. Jones and A. M. Flynn, "Mobile Robots", A.K. Peters, Wellesley MA, 1993.

[Mondada-et-al93]   F. Mondada, E. Franzi and P. Ienne, "Mobile robot miniaturisation: A tool for investigation in control algorithms", in Proceedings of the Third International Symposium on Expeirmental Robotics, Kioto Japan, 1993.

[Newell82]   A. Newell, "The Knowledge Level," Artificial Intelligence, 18, 87-127, 1982.

[Sommaruga93]   L. Sommaruga,: "Cooperative Heuristics for Autonomous Agents: an Artificial Intelligence Perspective," PhD Thesis, Nottingham University (UK), June 1993.

[Sommaruga&94]   Sommaruga, L. and Shadbolt, N. (1994), The Cooperative
                   Heuristics Approach for Autonomous Agents, in *Cooperative
                   Knowledge Base Systems - CKBS'94 Conference, Keele, UK,*
                   June 1994.

[Sommaruga-et-al95] L. Sommaruga, N. Avouris, M. Van Liedekerke: "The
                   Evolution of the CooperA Platform", to appear (late 95) in
                   Foundations of Distributed Artificial Intelligence, G. O'Hare
                   and N. Jennings editors, Wiley Inter-Science, Sixth-Generation
                   Computer Technology Series.

[Sommaruga-et-al95b] L. Sommaruga, E.M. Montero Viñuela , N. Paton Carrasco:
                   "An Architecture for Autonomous Robots Simulation," Third
                   International Symposium on Intelligent Robotic Systems -
                   SIRS 95, Pisa (Italy), 10-14 July 1995.

[Wielinga-et-al92] B.J. Wielinga, A.TH. Schreiber and J.A. Breuker, "KADS: A
                   Modelling Approach to Knowledge Engineering - 1991 -,"
                   Knowledge Acquisition, 4, (1), 1992.

[Zadeh&69]   L.A. Zadeh and E. Polak, "System Theory," Mc Graw-Hill,
                   1969.

Motivation and Perception Mechanisms in Mobile Agents for Electronic Commerce

C. Munday, J. Dangedej, T. Cross, and D. Lukose

Distributed Artificial Intelligence Centre (DAIC)
Department of Mathematics, Statistics, and Computing Science
The University of New England, Armidale, 2351, N.S.W., AUSTRALIA

Email: {cmunday I jirapun I tcross I lukose}@neumann.une.edu.au
Tel.: +6 (0)67 73 2302   Fax.: +6 (0)67 73 3312

Abstract: In this paper, the authors describe a mobile agency and an extendable mobile agent architecture that are specifically designed for Electronic Commerce. These agents are composed using two categories of mechanism: Reflexive Mechanism; and Deliberative Mechanism. Reflexive Mechanisms run concurrently in the agent, and the agent has no way of consciously controlling or directing its functions. On the other hand, the functions of Deliberative Mechanisms are consciously directed by the agent. This paper describes two Reflexive Mechanisms used in building deliberate agents: *perception*; and *motivation* mechanisms. We explore the types of knowledge and information stored in the Long-term and Short-term memory of the agent, how this knowledge and information are used by these mechanisms to perceive the external world, how the motives of the agent are revised, and finally, how motivations are generated.

Keywords: Motivation, Perception, Mobile Agent, Electronic Commerce, Conceptual Graphs, Executable Conceptual Structures, KQML, KIF.

1. Introduction

Motivation, emotion, and *cognition* are ideas much developed in psychology. Motivation refers to the meaning and purpose of behaviour. Emotion refers to experiential and psychophysiological phenomena that accompany motivational processes. Cognition refers to organismic activity, not directly observable, which translates motivation and emotion into observable behaviour [23]. An array of mechanisms are required by an autonomous agent to function in cyberspace (i.e., its world). Some of these mechanisms fall into the category of *reflexive* mechanisms, while others are known as *deliberative* mechanisms. These reflexive processes effect the agent's behaviour in such a way that the agent's actions almost always will be in response to a certain combination of *perception, emotion*, and *motivation*. On the other hand, the deliberative processes like *prediction, self-reflection, reasoning, goal creation*, and *planning* take into account the agent's perception of the external world, its current motivation and emotional state. This inter-dependency is what makes the agent behave in a rational manner in the cyberspace.

The conjecture in this paper is that processes like motivation, emotion, and perception are reflexive in nature, while cognition is deliberative. Perception and motivation are two types of reflexive processes. The agent (be it human or artificial) has no way of consciously directing these processes. The agent has to perceive the external world in which it lives and to interact with it. For it to be able to survive and carry out its functions in the cyberspace, it must have the capability to perceive changes in its world, understand messages from other agents, and also be able to perceive the effects of its own actions on the world and on other agents. There are several other attributes which an agent must have. In this paper, we address only the perception and motivation mechanisms, the knowledge and information requisite to their function, and these processes. We review a Mobile Agent architecture that is based on conceptual structures, and describe in details how conceptual graphs are used in the implementation of the perception and motivation mechanisms. We argue that an artificial agent need not have the emotional attributes to function in cyberspace in order to carry out electronic commerce.

The outline of this paper is as follows: In Section 2, we describe electronic commerce and the advantages gained in using a mobile agent for this purpose. This sets the stage for specifying the requirements of the mobile agency and the architecture of the mobile agents, which are described in Section 3. Sections 4 and 5 address perception and motivation processes, their requisite information and knowledge, and their mechanics, respectively. Finally, Section 6 concludes this paper by outlining future research directions.

2. Electronic Commerce

Electronic Commerce (EC) is used to described business process that functions electronically. Some contemporary examples include: ATMs; Electronic funds transfers; and most current banking transactions. EC replaces repetitive and mundane tasks with an automatic or electronic process. One technology that can be effectively used in EC is the *Mobile Agent*. A mobile agent is a computer program that can travel from its source to a number of different servers until it completes its given task [1]. There are several advantages in using mobile agents, in particular, these agents can reduce traffic in public networks [1] [9]. For example, when a user requires only a paragraph of text, rather than a user down loading the whole file across the network, a mobile agent can go to the server where the document is kept and search through the relevant parts of the document, only sending back the parts that the user requires. This approach will avoid transferring large data files across the network. In addition to reducing network traffic, other advantages of using mobile agents in EC include: saving operation costs, minimising intermediaries between producer and consumer, and enabling the professional skills of experts to be utilised by other users. An area of EC in which an *intelligent* mobile agent can be used is in Stock Market Trading. A human broker is a person (or organisation) who acts as a middleman between the stock seller and buyer. If a client (a person or an organisation) wants to buy a certain amount of stock, they must buy it via their broker, and their broker will then buy that requested amount of stock from someone who wants to sell that particular stock. By so doing, the broker will receive a certain amount of commission from either buyer or seller, or both.

One of the main motivations of the human broker is to maximise his/her commissions. To do this, the broker has to maintain the currency of his/her knowledge, as well as being a skilled stock analysts. The broker must have a large amount of digested information on stock trading [11] [22]. Human brokers need time to digest these continuously changing information. This limits their performance which directly effects client satisfaction. This indirectly effects the profitability of the human broker. Mobile agents can take the place of the current human brokers. Similar to a human broker, a mobile agent brokers not only trade stocks for their clients, but they also advice clients on current status of the stock market. Similar to the human broker, the motivation which drives mobile agent brokers to give the best service to their clients is the profit gain (commission). In addition, these agents must also maintain a reasonable number of clients to ensure maximum commission, but at the same time must ensure client satisfaction. As in human brokers, the motives have priority, and the behaviour of the agent is driven by these collective motivations. Some of the advantages of using a mobile agent broker are its ability to function 24 hours a day, as well as trading in several stock markets around the world. Also, mobile agent broker can service clients from around the world, and these clients can trade stocks through this agent in different markets around the world, as opposed to the current practice. Also, while trading in one market, these agents could keep track of stock price changes in other markets and advise its client on these changes, and on certain occasions, suspend its activity in one market, travel to another, do some trading and return to continue its activities. Sophisticated mobile agent broker could create a clone or a slave agent, assign it a specific task and send it away to carry out a task on its behalf at some other stock market. Clients registered with mobile agent have access to the most up-to-date information. In addition, the stock broking company can maintain several of these agents trading in different markets that are open at the same time. These agents can cooperate with each other to maximise returns in terms of monetary value as well as client satisfaction. Mobile agent technology can be effectively used to prevent illegal tradings (use of client's money to buy/sell stocks using bogus names). These types of crimes can be prevented by implementing mobile agent brokers which perform their tasks only in a legal way. Share markets around the world should only accepts mobile agent brokers that comply with certain legal practices that are consistent with the laws of the country in which these agents want to trade. Generally, human brokers are almost always only interested in *power* clients who have the capacity to trade large amounts of stock. They provide special services for these types of clients (eg., produce current reports each day or personally contact them by phone when some important event may effect their stock price). Smaller clients can only get information from public media such as television, radio, or newspaper which are quite slow at disseminating this information. Mobile agents could be designed to allow everyone to have equal access to the agent, and to the most up-to-date information. All clients will have the same opportunity to trade in international markets on equal grounds.

In a nutshell, clients have access to accurate, wider range, and high quality data because mobility allows the agent to be closed to the actual source of data to maintain currency of the information requisite to itself as well as its clients. Further, the mobile agent brokers can autonomously trade on behalf of its client without explicit authorisation at each time, given that the agent is conducting its trade within the specification outlined by its client (i.e., in terms of the amount and type of stock to trade, as well as the on the total monetary limit on the transactions). This new technology is beneficial financially and it generally improves the quality of life for all. Section 3 describes a model of a mobile agency, and an extendable architecture of mobile agents suitable for electronic commerce.

3. Mobile Agent Architecture

3.1. Mobile Agent Environment

Agent Server: This is the point where the agent arrives, performs its tasks and then departs to some other agent server. Our agent server architecture is made up of an IN-QUEUE, and OUT-QUEUE, service rooms and a stationary gatekeeper agent. The gatekeeper is responsible for managing the arrival and departure of the mobile agents. This includes the authentication of incoming agents, registration and allocation of a service room and the bundling and dispatch of agents leaving the server. Figure 1 depicts the architecture of the Agent Server. As mobile agents can be large and complex, it is important to identify techniques which can reduce network traffic. In our architecture, a mobile agent only needs to travel to a server with all its components once. When an agent arrives at a server for the first time, it is authenticated by the gatekeeper and allocated a service room. On subsequent visits, the same service room will be assigned to the agent. This allows the agent to deposit many of its bulky components in this room (ie., particularly site specific components /data) and know that these components will be available on subsequent visits. This facilitate a reduction in network traffic by reducing the number of components an agent needs to carry frequently visited sites. The service room can also provide the agent with an access point for the local services provided by the agent server. This allows for the possibility of utilising different categories of service rooms. Each category can provide different levels of

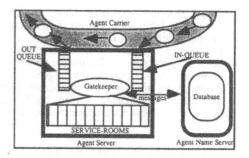

Figure 1: Agent Server

local services depending on factors such as the level of *trust* or access the local server wants to provide to the agent. The service rooms can also provide a set of default components which an agent may utilise. Agents have the choice of supplying their own components or utilising the default components provided by the server. The IN-QUEUE is where the incoming agents queue up for authentication before being allowed into their room, or before being allocated a new room for first time visitors. First time visitors have to register their components and request access to necessary local services and components with the gatekeeper. If their registration and service requests are approved by the gatekeeper, the agent is assigned an appropriate room. Once their service room has been allocated, the agent can instantiate itself and start up its components in order to carry out its tasks. The OUT-QUEUE is where the agent queues up to be transported to its next destination. Before departure, the mobile agent communicates to the gatekeeper it's new destination and lets the gatekeeper know the level of encryption needed. The gatekeeper performs the encryption process and dispatches the agent onto the network.

Agent Name Server: An Agent Name Server (ANS) functions similar to the Domain Name Servers used in the Internet. The ANS is necessary because during transportation, the mobile agent is not active and cannot respond to messages. Furthermore, it is not possible to predict the communications port address that an agent will be assigned on arrival at a new site. Therefore, as mobile agents move around in cyberspace, the ANS needs to be updated to indicate the status of the agent (stationary or in transit). For stationary agents, the name server will record the agents current communications port address. When an agent is in transit, the server will record the proposed destination. The ANS also provides a messaging service. Messages sent to an agent in transit can be stored by the ANS. On establishing a new communications port at it's destination, the ANS can forward to it the stored messages. Another service provided by the ANS is an audit trail of an agents movements. This is important for agents which deal in sensitive information or conduct commercial activities, such as a stock market broker.

Gatekeeper: This stationary agent resides in the Agent Server. It is responsible for the decryption, authentication and service room assignments for incoming agents and the encryption and dispatch of outgoing agents. The gatekeeper is also responsible for relaying agent status information to the ANS.

Agent Languages: Three languages are involved in our architecture. This first is the Tool Command Language (Tcl) [18], a scripting language which is used to implement the agents. Tcl is used because it is an interpreted language which can be easily extended, can provide a clean interface to other languages and has the potential to allow Agent Servers to verify and validate the agent's code. Furthermore, the design and implementation of Tcl supports the modification and extension of the Tcl interpreter. This enables Agent Server providers to restrict or extend the services that are offered by their sites. The second language is the Knowledge Query and Manipulation Language (KQML) [5], and the Knowledge Interchange Format (KIF) [6], used for agents communication. KQML is based on speech act theory. It provides the agent with standard syntax for messages, and a set of performatives to define different types of messages. KQML was selected because it has been designed specifically for the purpose of exchanging and manipulating knowledge between systems. Providing our agents with KQML communication should extend their flexibility by increasing the number and types of agents and sites they can communicate with. The third language is based on Conceptual Graphs [24] and Executable Conceptual Graphs [14]. These are used for representing the knowledge used in problem solving.

3.2. Extendable Agent Architecture

We adopted an extendable architecture for building our autonomous agents. This approach enables us to build agents that range from being purely reactive, highly deliberative, or a hybrid architecture. Our agent architecture has two types of components: *Physiological Components* and *Processor Components*. For example, the Short-Term Memory (STM) and the Long-Term Memory (LTM) are Physiological Components; whereas Processor Components are mechanisms such as perception; motivation; prediction; self-reflexion; goal creation, identification and evaluation; planning; plan identification and plan execution. Some of these mechanisms are reflexive, while others are deliberative. This list is by no means exhaustive. As our understanding of human cognition increases, we may be able to incorporate new components into our agent architecture. Each component is built using the *plugs* and *sockets* metaphor. Plugs are the input points, while the sockets are the output points. Each component may have a number of sockets and plugs. Each socket and plug has different configurations, and they are represented as conceptual graphs [24]. We use the symbol '«' between two conceptual graphs to indicate that one graph is a specialisation of the other [24]. For example, $g_1 « g_2$, indicates that g_1 is a more specialised graph of g_2. The following definitions clarifies the above descriptions.

Definition 3.2.1: For each component C_k (physiological or processor) of agent A with n components, there is a set of sockets $S_{1..j}$, such that there are no two sockets with the same configuration, no socket is subsumed by another: $S_{1..j} \in socket(C_k)$, where $1 \le k \le n$, $S_e \ne S_f$, and $not(S_e « S_f)$, where $S_e \in \{S_{1..j}\}$ and $S_f \in \{S_{1..j}\}$.

Definition 3.2.2: For each component C_k (physiological or processor) of agent A with n components, there is a set of plugs $P_{1..n}$ such that there are no two plugs with the same configuration, or one plug is subsumed by another: $P_{1..t} \in plug(C_k)$, where $1 \le k \le n$, $P_r \ne P_q$, and $not(P_r « P_q)$, where $P_r \in \{P_{1..t}\}$ and $P_q \in \{P_{1..t}\}$.

Definition 3.2.3: An agent is made up of n components, coupled together using sockets and plugs from each of the components. Let A be the agent, and $C_{1..n}$ be the components used to build A: $A \Longleftrightarrow C_{1..n}$, where $C_k = \{S_{1..i}\} \cup \{P_{1..t}\}$ according to Definition 3.2.1 and 3.2.2.

Definition 3.2.4: Two components C_k and C_l, where $C_k \ne C_l$, can be coupled iff the following conditions hold: $S_j \in socket(C_k)$ and $P_i \in plug(C_l)$ such that P_i can be projected onto S_j according to Definition 4.1.

Definition 3.2.5: The compatible projection of a plug P_i of a component C_k to the sockets $S_{1..j}$ of component C_l is performed according the following: Obtain all projection of plug P_i onto sockets $S_{1..j}$. Selected among the projected graphs, the graph that has the smallest *semantic distance* [24] between itself and P_i. If there are more than one resultant graph with the same semantic distance, then arbitrarily choose one.

Figure 2 depicts the plugs and sockets of the Physiological Components and the Motivation Mechanism, while Figure 3 shows how the sockets and plugs of these components are used to couple them together. Each of the these components have both plugs and sockets to enable input/output to take place. For example, the STM requires both plugs and sockets, so that other mechanisms like the Motivation Mechanism can plug onto one of its sockets to retrieve information stored in STM. On the other hand it must also have its own plugs that can be used to obtain information from other components so that its memory could be revised. Note that neither the STM or LTM have the consciousness to retrieve information from other components to revise themselves. In fact, it is the processor components that will retrieve information for processing and also have the capability to revise the content of STM and LTM. Note that (almost always), only the content of the STM will be revised. Only agents that have the ability to learn new heuristics may modify their LTM.

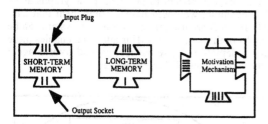
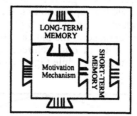

Figure 2: Sockets and Plugs of Components   Figure 3: Closely coupled components

Extendability implies that the components that make up a particular *type* of agents are modularised, and a particular agent could be assembled with only a subset of these components which are necessary for it to carry out its functions. Say for example, component A and component B have two plugs and two sockets each. Let g_{pa1} and g_{pa2} represent the two plugs for component A, and g_{sa1} and g_{sa2} represent the two sockets of component A. Similarly, we have g_{pb1} and g_{pb2} representing the plugs, and g_{sb1} and g_{sb2} representing the sockets of component B. When we want to couple these two components, we perform a projection of these graphs [24] to find the subgraph of the plug which is embedded in the socket graph. The choice of plugs or sockets of a component to be used in the projection operation is determined by whether the component wants to send information to the other, or receive information from the other component. Say, for example, if component A wants to send some information to component B, then the plugs of A and the sockets of B are used in the projection operation. The amount of information flow can be restricted by the choice of plugs and sockets that are allowed to be used in a projection operation. The information that will flow between these two components is determined by the projected image of the plug. The principles of projection is outlined in Section 4. Let us consider a

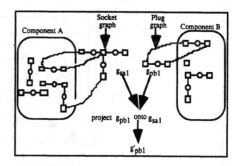

Figure 4. Sockets and Plugs of
Components A and B

simple example: Consider components A and B as depicted in Figure 4. There is a socket graph for component A and a plug graph for component B. Note that each component may have more than one socket and plug graph. Further assume that component B wants information pertaining to plug g_{pb1} from component A. Thus, coupling these two components would imply that graph g_{pb1} is projected onto graph g_{sa1}. The projected graph g'_{pb1} is the image of g_{pb1} in g_{sa1}. This projected graph becomes the "link" that couples these two components. Note that the gray lines between graphs in component A and B connect to their respective socket and plug graphs. This is the coreference link between the concepts of the graphs in the component and the concept of their plug and socket graphs. These coreference links must be transferred to the projected graph. The coreference link in the projected graph enables the two components to update their information content in a consistent manner. There are several advantages resulting from this form of coupling. Firstly, searching for the correct knowledge structure to update or to read is minimised as direct connections to these structures are maintained by the coreference links. Secondly, we don't have to explicitly write the code to interface between our components. Using the plug and socket graphs, we are able to dynamically link the component. It is much simpler to modify the graphs than to write code when the components are modified. Flexibility is the major advantage. It is not necessary for each component to use conceptual graphs as its knowledge representation scheme. For example, we could have a component that uses productions as in CLIPS [7]. The predicates that are used in the production system can be coreferenced to the plug and socket graphs of this component. This approach enables more flexibility in our architecture. Figure 5 depicts reactive, deliberate and hybrid agent architectures, while Figure 6 outlines the components of a deliberate agent.

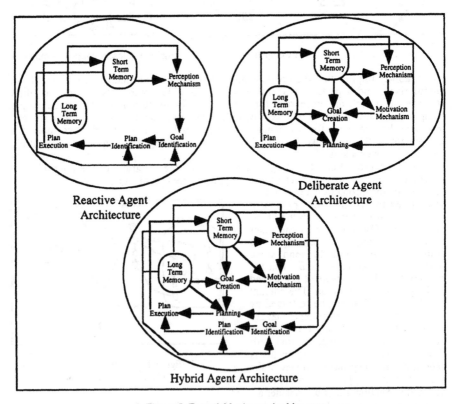

Figure 5. Extendable Agent Architecture

4. Perception Mechanism

The agents communicate with the external world (and with other agents and information servers) using the agent communication language KQML and the Knowledge Interchange Format (KIF). Here, we are only concerned with the perception of incoming messages from other agents. When the agent wants to know about certain aspects of the external world, it either sends a request to a server agent or other mobile or stationary agent for information. The other agents or servers reply by sending back appropriate messages. Perception is a process by which the agent is able to understand the external world in terms of its own

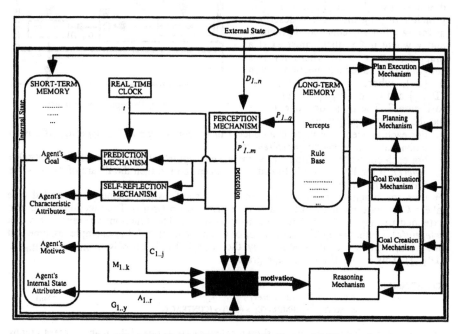

Figure 6. A Detailed Architecture of a Deliberate Mobile Agent

percepts (i.e. knowledge). The percepts can be thought of as skeleton structures that can be used by the perception mechanism of the agent to impose structure on incoming messages and thereby assist the agent in interpreting the information. These percepts can be represented by conceptual graphs. The percepts of an agent are stored in its LTM and are projected onto the incoming message graphs to produce its perception of the message. The projection of the percept graph onto the message graph results in a third graph which represents a specialisation of the skeletal percept graph. This graph is then used by various other components in the agent to reason, predict, self-reflect and revise its motivation. A similar message handling-mechanism is described in Lukose [15]. Figure 7 depicts the perception mechanism in our agent. For example, consider an incoming KQML message. The KQML MESSAGE HANDLER will strip out the message and obtain the contents. The message content may be represented by a conjunction of knowledge structures. For example, let us denote them as $D_{1..n}$, where $n \geq 1$. These chunks may be in the form of conceptual graphs or any other form of representation scheme. If it is in another form of representation scheme, these chunks arepassed on to the KIF PROCESSOR which transforms them into conceptual graphs. The transformed message graphs are denoted as $D'_{1..r}$. The PROJECT component will retrieve the percepts (denoted as $P_{1..q}$) from LTM and perform projection [24] of the percepts onto the message graphs $D'_{1..r}$ to produce the interpretation of the incoming message denoted by $P'_{1..m}$ in Figure 7. The principle of projection is best described as follows. For any conceptual graphs u and v where $u \leq v$, there must exist a mapping $\pi: v \rightarrow u$, where πv is a subgraph of u called a projection of v in u [24], where $v \in \{P_{1..q}\}$, $u \in \{D'_{1..r}\}$ and $\pi v \in \{P'_{1..m}\}$) in reference to Figure 7. The projection operator π has the following properties:

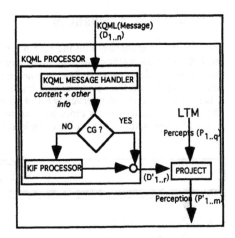

Figure 7. Perception Mechanism

- For each concept c in v, πc is a concept in πv where $type(\pi c) \leq type(\pi v)$. If c is an individual, then $referent(c) = referent(\pi c)$.
- For each conceptual relation r in v, πr is a conceptual relation in πv where $type(\pi r) = type(r)$. If the ith arc of r is linked to a concept c in v, then the ith arc of πr must be linked to πc in πv.

Further, the function $type$ maps concepts into a set T, whose elements are called type labels. Concept c and d are of the same type iff $type(c) = type(d)$. The function $type$ is also extended to map conceptual relation to type labels. The function $referent$ states that $referent(c)$ is either an individual marker or a generic marker *. For example, assume that one of the graph that is produced by the KQML PROCESSOR is g_{m1} (i.e., $g_{m1} \in \{D'_{1..r}\}$). This is one of the message graph.

```
[CLIENT: Nerida] -
        -> (LOC) -> [PLACE: Sydney]                              - g_m1
        <- (AGNT) <- [TERMINATE] -> (OBJ) -> [CONTRACT]
```

Further, assume that a graph pg_1 as shown below exists as one of the many percept graphs in the agent's LTM.

```
[PERSON] <- (AGNT)<- [TERMINATE] ->(OBJ)-> [ENTITY]              - pg_1
```

The PROJECT part of the perception mechanism will perform a projection of pg_1 onto g_{m1}, and an image of pg_1, written as pg'_1 will be produced. This graph represent the perception of the agent of the message just received.

```
[CLIENT: Nerida] <- (AGNT) <- [TERMINATE] -> (OBJ) -> [CONTRACT]    - pg'_1
```

5. Motivation Mechanism

The Motivation Mechanism of an agent is a reflexive component which takes the perception, and modifies the internal state attributes of the agent, revises the values of the motives of the agent, and generates appropriate motivation to enable the agent to behave rationally in the cyberspace. There are four types of information stored in the STM that are used by the Motivation Mechanism. These types of information are: Agent's goal[s],

Agent's Characteristic Attributes, Agent's Internal State Attributes; and, Agent's Motives. Each are represented as conceptual graphs. Below, we describe each of these types of information, and in Section 5.2 we outline the motivation mechanism and describe its functions.

5.1. Information used by Motivation Mechanism

Agent's Goals: A Goal is a state that the system can achieve or prevent through actions (i.e., invoking planning and execution processes). Agents react to a situation by creating a goal (i.e., reactive goal) to manage this situation (triggered by a change in belief), and by planning a strategy to satisfy the goal. Almost all of the contemporary literature describe the creation of reactive goals within Autonomous Agents [26]. The range of goal types may be classified as: reactive; proactive; achievement goals; prevention goals; and maintenance goals [25]. Recently, considerable intellectual work has been devoted to analysing the creation, use, and effects of proactive goals [4] [16] [17]. The *information attitudes* and *pro-attitudes* of our agent is very much in the line of BDI-architecture (Belief, Desire, Intention) of Rao and Georgeff [19] [20]. The need for an Autonomous Agent to be able to predict the future is paramount for the achievement and/or prevention of certain states through goal identification, planning, and plan execution. This implies that the information processing mechanisms within an autonomous agent will have to consider goal types other then reactive goals. Our agent architecture and information processing mechanisms are designed with this in mind.

Agent's Characteristics Attributes: A sophisticated agent interacting in a rational manner in the cyberspace needs a set of characteristics attributes that defines its behaviour, and the values of these attributes contribute to the motivations of the agent, goal creation, and eventually planning. Examples of such attributes in a mobile agent broker are: ability, effort, persistence, and profit making policies related to clientele and the client satisfaction policies. An agent who perceives the external environment, can perform the deliberate process of self-reflection to revise the values of these attributes in the hope that the effects of its behaviour in the cyberspace will improve its current predicament. The motivation mechanism does not modify the values of these attribute. In our current architecture, values of these attributes can only be modified by the deliberative mechanism called SELF-REFLECTION MECHANISM. This mechanism interacts with the PREDICTION MECHANISM, which in turn interacts with the GOAL CREATION mechanism, to create proactive goals to evaluate different scenarios. Descriptions of these deliberate mechanisms are beyond the scope of this paper. The MOTIVATION MECHANISM only utilises the values of the agent's characteristics attributes to revise the values of its motives.

Agent's Internal State Attributes: The internal state attributes represent the perceived changes in the agent's environment. Examples of these attributes may include: prices of various commodities in different market places; belief about its own abilities; beliefs on policies of other broker agents; profit level; shares traded per-day; and number of clients. The agent's internal state attributes are revised by a reflexive component called the INTERNAL STATE REVISOR (c.f. Figure 8). The value of the agent's internal state variables are used by the MOTIVATION GENERATOR (c.f. Figure 8) for triggering appropriate motivation[s] for the agent. For example, if a mobile agent broker has the motive for maintaining a certain profit margin, then this motive may be influenced by the share prices of certain stock and/or the policies of other agents in the environment. Values of both of these internal state attributes will have an effect on the motivation to maintain a certain profit margin. The motivation will be increased if, for example, other agents change their policies in such a way that it leads to an adverse effect on the amount of money this agent is making. Based on these types of motivation[s], the agent has the ability to generate goals.

Agent's Motives: Motives are best defined as the *need* or *desire* that causes an agent to act[12]. Norman and Long [17] define motive as a variable that has influence on the decision-making process and may cause action. Motive, Motivation, Goals and Plans are inter-linked to each other. Motivation is the driving force towards achieving a goal. It is the measure of the importance of the goal to the agent at a certain time. In our architecture, motives are influenced by internal state variables. These variables are shared among different motives. The relative effect of each motive is determined by the collective weights of the internal variables that influence the motive. Each motive is assigned a *value attribute* and a *threshold*. The value can be either qualitative or quantitative. The value attribute is revised by the MOTIVE REVISOR based on the agent's perception of the world (c.f., Figure 9). When the value attribute of a motive passes its threshold, it will be triggered. At any one time there may be a set of motives that have passed their threshold values. All the motives in an agent are related to each other in terms of priorities. The priority of the motives to each other changes according to the current goal of the agent. The MOTIVE REVISOR takes the current goal of the agent into account when revising the values of the motives. Once a motive has been triggered it is inhibited from further influencing the agent's actions until either: (1) the goal that was generated as a result of the trigger is dropped; or (2) the goal has been planned and executed. A motive may be in one of two states: *SLEEPING* or *AWAKE*. A motive that is *SLEEPING* has just been triggered and the subsequent goal has not been achieved or has not been dropped. A motive that is *AWAKE* has the possibility of being triggered due to sufficient motivating factors. The inter-relationships between agent's goals, characteristic attributes, internal state values and motives are represented in the productions stored in LTM. Changes to the motive priorities are also embedded in the productions in LTM.

5.2. Motivation Mechanism

The Motivation Mechanism is made up of three components. They are: Internal State Revisor, Motive Revisor, and Motivation Generator. Figure 8 depicts the architecture of the Motivation Mechanism. These three components are only related to each other by the information that they utilise. From all other points of view, they are independent of each other, and they function in parallel (concurrent on a sequential machine). The information that they utilise comes from the STM and LTM. The STM contain agent motives, agent characteristics, and agent's internal state variables. The information from LTM that is utilised by these three components is the production rules appropriate to each component. Each of these components are simple inference engines. Based on the content of their working memory, appropriate productions will be executed. We adopted the use of production systems simply because the productions are unordered, and every production in the entire system is potentially active at every moment. It is simply waiting, independent of the rest of the system, for the announcement that its conditions are satisfied [21]. Figure 9 depicts in more detail the MOTIVE REVISOR.

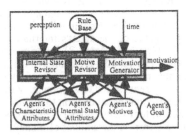

Figure 8. Motivation Mechanism

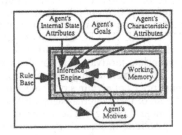

Figure 9. Motive Revisor

The productions stored in the LTM are in the form of conceptual rules, as shown below:

[LHS: <conjunctive conditions>] -> (THEN) -> [RHS: <actor graph>]

where <conjunctive conditions> are simply conceptual graphs separated by comma. An <actor graph> is a form of executable conceptual structures developed as an extension to the original conceptual graph theory to enable dynamic changes to conceptual structures [14]. An actor graph for a particular type is made up of the differentia graph (i.e., declarative component) for the type and the actor (i.e., procedural component) for activating the action represented by the type [13]. For example, assume that one of the internal state attributes is the clientele number, represented by the conceptual graphs g_1, and the perceived graph is g_2, as shown below. Further assume that there is a production p_1 as listed below in the agent's LTM.

[CLIENTELE] -> (QTY) -> [NUMBER:15]   · g_1

[CLIENT: Nerida] <-(AGNT) <-[TERMINATE]->(OBJ)->[CONTRACT]   · g_2

[LHS: [CLIENTELE] -> (QTY) -> [NUMBER:*y]   · p_1
            [CLIENT: *x]<-(AGNT) <-[TERMINATE]->(OBJ)->[CONTRACT]
] -> (THEN) ->
[RHS: [SUBTRACT]-
                ->(QTY)->[NUMBER:1]
                ->(FROM) -> [NUMBER:*y]
].

The above production p_1 simply states that the clientele quantity is represented by a variable *y, and if the agent perceived that a client identified by variable *x terminates the contract with the agent, then invoke the actor graph to subtract 1 from the clientele quantity. The actor graph for SUBTRACT has the procedure to subtract a number from another number. In this case, it will subtract 1 from the value represented by *y. Note that *y also indicates coreferencing between the two NUMBER concepts in p_1. When g_2 is received by the INTERNAL STATE REVISOR, it will retrieve the appropriate internal state attribute (i.e., in this case graph g_1) and load both g_1 and g_2 into its working memory. It then selects all productions that has its conditions satisfied. In this example, production p_1 may be the candidate. Production p_1 will then get instantiated with the information that is currently in the INTERNAL STATE REVISOR's working memory. This instantiation will generate a production p'_1 as shown below:

[LHS: [CLIENTELE] -> (QTY) -> [NUMBER:15]   · p'_1
            [CLIENT: Nerida]<-(AGNT) <-[TERMINATE]->(OBJ)->[CONTRACT]
] -> (THEN) ->
[RHS: [SUBTRACT]-
                ->(QTY)->[NUMBER:1]
                ->(FROM) -> [NUMBER:15]
].

Execution of the actor graph SUBTRACT will reduce the clientele number by one. As both the NUMBER concepts are coreferenced, any changes to the one NUMBER concept will be reflected in the other NUMBER concept. The resulting value of the clientele internal state variable will be as shown in graph g'_1.

[CLIENTELE] -> (QTY) -> [NUMBER:14]   · g'_1

6. Conclusion

In this paper, we outlined a mobile agency suitable for electronic commerce, which requires two types of agents: Mobile Agents; and, Gatekeeper. Our architecture also relies on an Agent Server to provide services to our mobile agents and an Agent Name Server which facilitates agent tracking and communication. We outlined an extendable agent architecture based on the metaphor of plugs and sockets and utilises the conceptual structures formalism and the projection operator. Further, we reviewed the architecture of a deliberate agent, and outlined the perception and motivation mechanism. These are the two reflexive mechanisms in our agent architecture. Finally, we outlined (with simple examples using conceptual graphs) how the perception and motivation mechanisms work.

One of the main advantage of a mobile agency is a reduction in network traffic. The mobile agent architecture outlined in this paper demonstrates techniques for reducing network load by reducing the size of these agents. We could further reduce the size of the mobile agents by enforcing each Agent Server to maintain a set of core components necessary for stock trading. All broker agents will have access to all of these core components in its Service-Room.

Thus far, each component of the mobile agent is being developed separately. These components have to be merged to carry out extensive experiments in the domain of stock trading. One of the main areas of work that is not covered in this paper is Agent Security. Sound solutions to agent security issues like origin authentication, data integrity, access/itinerary control, agent's privacy, and privacy and integrity of gathered information are important issues which will determine the successful application of mobile agent technology in electronic commerce. This is one of the main areas of future research. The other area of research that has been neglected thus far is Mobile Agent Modelling. Formal methods for the knowledge level modelling of an agent is fundamental for successfully building mobile agents. Formal methods are necessary to fulfil the agent verification requirements so as to ensure that it functions according to specification. Some pioneering work in this area has been done by Dieng [2], Glaser, et. al. [8], and Huang, et al. [10] using the KADS approach, while Dunin-Keplicz, et. al. [3] have used the DESIRE modelling approach.

Mobile agents are a reality at the moment. Their successful use in electronic commerce will certainly be a reality in the near future. Many procedures need to be refined in order to realise this emergent technology. This paper has addressed a few of the issues. Many issues still remain to be resolved.

7. Acknowledgments

The research outlined in this paper is carried out with the following financial support: Australian Postgraduate Scholarship (APA), UNE Research Initiative Scheme, Internal Research Grant from the Department of Mathematics, Statistics, and Computing Science at UNE, and the Assumption University PhD Scholarship. We also acknowledge the help of Meg Vivers, Arno Puder and the anonymous referees who have helped to improve the quality of this paper.

8. References

1.   Chess, D., Grosof, B., Harrison, C., Levine, D., Parris, C., and Tsudik, G. Itinerant Agents for Mobile Computing, *IBM Research Report*, IBM Zurich Research Laboratory, Switzerland, 1995.

2. Dieng, R. Specifying a Cooperative System through Agent-Based Knowledge Acquisition, in Proceedings of the 9th Banff Knowledge Accquisition For Knowledge-Based Systems Workshop, Banff Conference Centre, Banff, Alberta, Canada, February 26 - March 3, 1995, Paper No: 20.

3. Dunin-Keplicz, and B., Treur, J. Modelling Reasoning and Acting Agents, in Proceedings of the 9th Banff Knowledge Accquisition For Knowledge-Based Systems Workshop, Banff Conference Centre, Banff, Alberta, Canada, February 26 - March 3, 1995, Paper No: 22.

4. Ekdahl, B., Astor, E., and Davidsson, P. Towards Anticipatory Agents, *Proceedings of ECAI-94 Workshop on Agent Theories, Architectures, and Languages*, Lecture Notes in Artificial Intelligence, M.J. Wooldridge and N.R. Jennings (Eds.), No. 890, Springer-Verlag, Germany, 1994.

5. Finin, T., Weber, J., Wiederhold, G., Genesereth, M., Fritzson, R., McGuire, J., Shapiro, S., and Beck, C. Specification of the KQML Agent-Communication Language, 1993. http://retriever.cs.umbc.edu:80/kqml/

6. Genesereth, M.R., and Fikes, R.E., (1992). Knowledge Interchange Format Version 3.0 Reference Manual, Technical Report Logic-92-1, Stanford University, USA.

7. Giarratano, J.C. *CLIPS User's Guide*, Software Technology Branch, Information Systems Directorate, Lyndon B. Johnson Space Centre, NASA, USA, 1991.

8. Glaser, N., Haton, M-C., and Haton, J-P. Models and Knowledge Acquisition Cycles for Multi-Agent Systems, in Proceedings of the 9th Banff Knowledge Accquisition For Knowledge-Based Systems Workshop, Banff Conference Centre, Banff, Alberta, Canada, February 26 - March 3, 1995, Paper No: 40.

9. Harrison, C.G., Chess, D.M., and Kershenbaum, A. Mobile Agent: Are They Good Idea?, *Research Report*, IBM Research Division, T.Y. Watson Research Center. March, 1995.

10. Huang, J., Jennings, N.R., and Fox, J. An Agent Architecture for Distributed Medical Care, *Proceedings of ECAI-94 Workshop on Agent Theories, Architectures, and Languages*, Lecture Notes in Artificial Intelligence, M.J. Wooldridge and N.R. Jennings (Eds.), No. 890, Springer-Verlag, Germany, 1994.

11. Jessup, P.F. *Competing for Stock Market Profits*, John Wiley & Sons, Inc., 1974.

12. Luck, M., and d'Inverno, M. Engagement and Cooperation in Motivated Agent Modelling, *Proceedings of the First Australian Workshop on Distributed Artificial Intelligence*, 1995. (in this volume)

13. Lukose, D., and Garner, B.J. Actor Graphs: A Novel Executable Conceptual Structure, *Proceedings of the IJCAI Workshop on Objects and Artificial Intelligence*, Sydney, Australia, August 25, 1991.

14. Lukose, D., Executable Conceptual Structure, *Proceedings of the International Conference On Conceptual Structures, Theory and Applications*, August 4-7, 1993, Quebec City, Canada.

15. Lukose, D. Projection Based Invocation Mechanism for Actor Graphs. In *Proceedings of the 2nd Australia and New Zealand Conference on Intelligent Information Systems*, Brisbane, Queensland, Australia, 1994.

16. Moffat, D., and Frijda, N. Where There's a *Will* There's an Agent, *Proceedings of ECAI-94 Workshop on Agent Theories, Architectures, and Languages*, Lecture Notes in Artificial Intelligence, M.J. Wooldridge and N.R. Jennings (Eds.), No. 890, Springer-Verlag, Germany, 1994.

17. Norman, T.J., and Long, D. Goal Creation in Motivated Agents, *Proceedings of ECAI-94 Workshop on Agent Theories, Architectures, and Languages*, Lecture Notes in Artificial Intelligence, M.J. Wooldridge and N.R. Jennings (Eds.), No. 890, Springer-Verlag, Germany, 1994.

18. Ousterhout, J.K. *Tcl and the Tk Toolkit*, Addison-Wesley, 1994.

19. Rao, A.S. Integrated Agent Architecture: Execution and Recognition of Mental-States, *Proceedings of the First Australian Workshop on Distributed Artificial Intelligence*, 1995. (in this volume)

20. Rao, A.S., and Georgeff, M.P. Modeling rational agents within a BDI-architecture, in R. Fikes, and E. Sandewall (Eds.), Proceedings of Knowledge Representation and Reasoning (KR&R-91), pp. 473-484, Morgan Kaufmann, April, 1991.

21. Simon, H.R. The Bottleneck of Attention: Connecting Thought with Motivation, In *Integrative Views of Motivation, Cognition, and Emotion*, Volume 41 of the Nebraska Symposium on Motivation, William D. Spaulding (Ed.), University of Nebraska Press, Lincon, USA, 1994.

22. Stanley, S.R. *International Financial Market Intergration*, Brasil Blackwell Ltd., 1993.

23. Spaulding, W.D. *Integrative Views of Motivation, Cognition, and Emotion*, Volume 41 of the Nebraska Symposium on Motivation, William D. Spaulding (Ed.), University of Nebraska Press, Lincon, USA, 1994.

24. Sowa, J.F. *Conceptual Structures: Information Processing in Mind and Machine*, Addison Wesley, Reading, Mass., 1984.

25. Wilensky, R. *Planning and Understanding: A Computational approach to human reasoning*, Addison-Wesley, 1983.

26. Wooldridge, M., and Jennings, N.R. Agent Theories, Architecture, and Languages: A Survey, *Proceedings of ECAI-94 Workshop on Agent Theories, Architectures, and Languages*, Lecture Notes in Artificial Intelligence, M.J. Wooldridge and N.R. Jennings (Eds.), No. 890, Springer-Verlag, Germany, 1994.

Integrated Agent Architecture: Execution and Recognition of Mental-States

Anand S. Rao

Australian Artificial Intelligence Institute
Level 6, 171 La Trobe Street
Melbourne, Victoria 3000
Australia

Phone: (+61 3) 9663-7922, Fax: (+61 3) 9663-7937
Email: anand@aaii.oz.au

Abstract. Recognizing the mental-state – the beliefs, desires, plans, and intentions – of other agents situated in the environment is an important part of intelligent activity. Doing this with limited resources and in a continuously changing environment, where agents are also changing their own mental-state, is a challenging task. Following the relative success of reactive planning, as opposed to classical planning, we introduce the notion of *reactive plan recognition*. We integrate reactive planning and reactive plan recognition and embed them within the framework of an agent's mental-state. This results in a powerful architecture for agents that can handle executions based on mental-states and recognition of the mental-states of other agents.

1 Introduction

Agents are computational entities that are situated in dynamic environments, acting to fulfil desired ends, and reacting to changing situations. They continuously receive perceptual input from a changing environment and, based on their mental-state, exhibit reactive or pro-active behaviour by performing certain actions. The components of a mental-state of an agent (or the lack of such an explicit representation) lead to different types of agent architectures [2, 4, 17]. An important subclass of agents can be viewed as having mental-states that comprise the beliefs, desires, plans, and intentions both of themselves and of others. While it is reasonable for us, as designers of multi-agent systems, to provide for an agent an initial set of beliefs, desires, and plans (based on which it forms its intentions), it is often desirable that such an agent be then able to *recognize* the beliefs, desires, and intentions of the other agents in its environment.

There has been considerable work in recent years on plan recognition [9, 10] that focuses on inferring the plans and intentions of other agents. However, most of this work treats plan recognition as the reverse process of classical planning, concerned with inferring plans in a bottom-up fashion from observations. Furthermore, this approach is not integrated with the overall mental-state model of the agent, and is not suitable for resource-bounded agents situated in a dynamic environment.

Over the past decade the focus of research in planning has shifted from classical planning to reactive or situated planning [4]. Reactive planning is based on two premises: (a) the environment in which an agent is situated is continuously changing; and (b) agents situated in such environments have limited resources. These have led to the development of various architectures and techniques for guiding the agent in its decision-making process, for making agents commit to their decisions as late as possible and, once committed, to stay committed as long as possible, within rational bounds.

Research in reactive planning has led to the redefinition of the notion of plans. Plans are used in two different contexts: (a) plans as abstract structures or recipes for executing certain actions and achieving certain states of the world; and (b) plans as complex mental attitudes, also called intentions, intertwined in a complex web of relationships with other mental attitudes such as beliefs and desires [11]. Plans as recipes *guide* a resource-bounded agent in its decision-making process, thereby short-circuiting the time-consuming search through a possible space of solutions as is done in classical planning. Agents, under their current set of beliefs, respond to their desires by committing to execute certain plan recipes. Such committed plan recipes are treated as intentions and are executed by the agent, one step at a time. Plans as mental attitudes *constrain* the agent in its future decision-making by committing it to previously made rational decisions. The entire process is called *mental-state-based plan execution*.

In the case of reactive recognition, we extend the above notion of plans as recipes to recognition plans that contain operators for *observing* the execution of primitive actions and *recognizing* the achievement of certain states of the world. Embedding this notion of plans within the mental-state of an agent results in intentions towards recognition and desires towards recognition. The execution of such recognition intentions leads to an agent recognizing the desires and beliefs of other agents in the environment. We call this process *mental-state recognition*.

Thus, recognition plans play the same role as performative plans in reactive planning, by guiding a resource-bounded agent in its recognition, thereby short-circuiting the time-consuming search through a possible space of explanations for an observation, as is done in classical recognition [9]. Recognition intentions play a similar role to intentions in reactive planning by constraining the recognition process and committing to previously made rational decisions in terms of what actions or events to observe.

Such reactive planning and recognition can be combined into an integrated architecture allowing the execution of plans to be determined by mental-state recognition, and conversely the recognition of mental-states to be determined by plans one executes. However, such an integrated reactive system is based on three important assumptions: (a) the agent has correct and complete knowledge of the plans that fulfil its desires; (b) the agent has correct and complete knowledge of the plans of other agents whose mental-state it is trying to recognize; and (c) under any given situation the agent has a relatively small set of plans (execution or recognition plans) that it is trying to execute or recognize. This does not mean that the total number of plans be small – rather, that the plans *specific to a given situation* be small in number.

Reactive planning is based on the first and third assumptions. Given that plans have to be pre-specified in a plan library (Assumption (a)) the number of plans that need to be considered is fairly small (Assumption (c)). Similarly, in the case of reactive recognition a correct and complete set of plans of executing agents is required (Assumption (b)) by the recognizing agent to constrain the recognition process (Assumption (c)). If the recognizer's model of the executing agent is incorrect or incomplete, the recognition would still occur but the recognition would turn out to be incorrect or incomplete. In other words, the recognition is as good as the model the recognizer has of the observer. If Assumption (c) is not satisfied then reactive planning and reactive recognition become as good (or as bad) as their classical counterparts.

Elsewhere [12] we presented a limited form of reactive recognition called *means-end recognition* and provided a model-theoretic and operational semantics for it. In this paper, we extend the means-end recognition process to reactive recognition by modifying an existing Belief-Desire-Intention (BDI) interpreter [15] used for reactive planning. Given that reactive recognition is taking place within the framework of an agent's other mental attitudes, we can also infer the mental-state of other agents, thereby providing a theory of mental-state recognition.

The primary contribution of this paper is in laying the foundations for an integrated architecture for rational resource-bounded agents that allows such agents to act based on their mental-states and recognize the mental-states of other agents in a continuously changing environment.

2 Integrated Agent Architecture

We consider agent architectures where the mental-state of an agent consists of attitudes such as *beliefs, desires, plans,* and *intentions* [2, 15]. The attitudes of beliefs and plans represent the informational state of an agent; the former captures the declarative state of the agent and its environment (which may include other agents); while the latter captures the procedural information, i.e., how to achieve certain desires under certain belief conditions. The attitudes of desire and intention capture the motivational and decision states of an agent.

There are three major processes that operate on the mental-state of an agent: means-end reasoning; deliberation; and reconsideration. *Means-end reasoning* involves the agent processing its perceptual input (from the external environment) and motivational inputs (from within the agent itself) to generate *options* that the agent can then further deliberate upon. Options are instantiated plans that are applicable to the current situation. In the integrated architecture the plans may involve executing and observing the execution of primitive plans or actions, as well as achieving certain states and recognizing the achievement of certain states.

The *deliberation* process decides on the *best* (given the current resources) option(s) the agent should commit to or intend achieving. Some variant of decision theory [5] can be used for this deliberation process. The deliberation process takes place within the background set of current beliefs, desires, and intentions. As intentions capture the decision state of the past they are used to eliminate those options that were considered and rejected in the past, thereby constraining the deliberation process. At this stage, the agent has to achieve an appropriate balance between execution options and recognition options (i.e., to act or to observe).

Reconsideration takes place before deliberation and involves the agent re-examining current intentions in the light of options generated by new opportunities in the environment, new inferences by the agent or changed desires. The deliberation process may result in the reconsidered intention surviving or being replaced by new options. With reactive recognition, the reconsideration process becomes more powerful. A performative intention can now be reconsidered based on newly recognised mental-states of other agents. The converse could also occur, with a recognition intention being reconsidered based on certain actions taken by the agent.

As discussed earlier, an important aspect of the architecture is the notion of plans and the role they play in guiding an agent's behaviour. Now we consider the abstract structure of a plan and its operational semantics. A plan has (a) a name, (b) an invocation condition that can trigger the plan, (c) a precondition that needs to be believed by the agent before the plan body is started, (d) a postcondition that is believed by the agent after the plan body is finished successfully, and (e) a plan body which is a rooted, directed, acyclic AND-OR graph with the edges labeled with certain plan expressions. A plan expression can be one of four forms: (i) an agent a executes a primitive plan e, denoted by $(a\ e)$; (ii) an agent a observes the execution of primitive plan e by b, denoted by (recog $a\ (b\ e)$); (iii) an agent a achieves a proposition α, denoted by (! $a\ \alpha$); and (iv) an agent a recognizes the achievement by agent b of proposition α, denoted by (recog $a\ (!\ b\ \alpha)$).

The operational semantics of executing a primitive plan and observing a primitive plan are straightforward; in the former case the agent executes the primitive plan immediately and in the latter case the agent waits (indefinitely or until time-out) until the primitive plan is executed by the executing agent. When an agent desires to achieve a certain proposition it selects *a plan* whose invocation condition matches its desire and whose preconditions are believed and intends the body of the plan. However, when an agent desires to recognize the achievement of a desire it selects *all plans* whose invocation condition matches its desire and whose preconditions are believed and intends the body of all such plans. In either case, the agent is said to have successfully fulfilled its desire if at least one plan successfully completes by reaching the END-node. Operationally, control moves from one node to the next if the plan expression labeling the arc is completed successfully. In the case of an OR-node, it is sufficient for the plan expression labelling one of the arcs to be executed successfully for the control to move to the next node; in the case of an AND-node all the plan expressions have to be executed successfully.

⟨trigger⟩ := BEL-ADD ⟨bel-value⟩ | BEL-REM ⟨bel-value⟩ |
　　　　　　DES-ADD ⟨des-value⟩ | DES-REM ⟨des-value⟩ |
　　　　　　INT-ADD ⟨int-value⟩ | INT-REM ⟨int-value⟩ |
　　　　　　ACT-SUC ⟨act-res-val⟩ | ACT-FAL ⟨act-res-val⟩ |
　　　　　　DES-SUC ⟨des-res-val⟩ | DES-FAL ⟨des-res-val⟩ |
　　　　　　PLN-SUC ⟨pln-res-val⟩ | PLN-FAL ⟨pln-res-val⟩
⟨bel-value⟩ := ⟨prop⟩
⟨des-value⟩ := ⟨mode⟩ ⟨agent⟩ ⟨pln-expr⟩
⟨act-res-val⟩, ⟨des-res-val⟩ := ⟨end⟩ ⟨means⟩ ⟨intention-tree⟩
⟨pln-res-val⟩ := ⟨means⟩ ⟨end⟩ ⟨intention-tree⟩
⟨int-value⟩ := ⟨end⟩ ⟨means⟩ ⟨intention-tree⟩ | ⟨means⟩ ⟨end⟩ ⟨intention-tree⟩
⟨mode⟩ := exec | recog
⟨option⟩ := ⟨trigger⟩ ⟨plan-name⟩
⟨end⟩ := DESIRE ⟨des-value⟩ | PRM-ACT ⟨prm-act-value⟩
⟨means⟩ := PLAN ⟨mode⟩ ⟨name⟩ ⟨invocation⟩ ⟨context⟩ ⟨effects⟩ ⟨current-node⟩
⟨end-tree⟩ := ⟨end⟩→⟨means-tree⟩*
⟨means-tree⟩ := ⟨means⟩→ ⟨end-tree⟩+
⟨intention-tree⟩ := ⟨end⟩→ ⟨means-tree⟩+

Fig. 1. Data Structures

3 BDI Interpreter

Complex propositional modal logics of beliefs, desires, intentions, and time are used to specify the behaviour of agents [3, 14, 13]. Most of these logics do not have complete axiomatizations and are not efficiently computable. This naturally raises doubts regarding the feasibility of building modal logic theorem provers as the reasoning component of such agents and more seriously the practicality of building agent-oriented systems based on the notions of beliefs, desires, and intentions.

While the specification and formalization of agents can be carried out in complex modal logics, we believe that to obtain practical systems, at least in the foreseeable future, one needs to treat the mental attitudes as data structures that satisfy certain properties, and build an interpreter that operates on these data structures. This is the approach taken in this paper.

An abstract BDI-interpreter for performing reactive plan execution within the context of the beliefs, desires (or goals), intentions, and plans of an agent was given elsewhere [15]. However, the details involved in the means-end reasoning of reactive planning were not discussed. In this section, we give a detailed presentation of the BDI-interpreter and extend it to reactive recognition, thereby integrating both reactive plan execution and recognition.

Before considering the details of the algorithms, we describe some of the data structures used by the BDI interpreter. The BNF notation for the above data structures is given in Figure 1.

A *trigger* is an internal event which consists of a *trigger type* and a *trigger*

```
procedure reactive-reasoner(P, M_S, trigger-list, M_T)
options := option-generator(P, M_S, trigger-list);
selected-options := deliberate(M_S, options);
M := update-mental-state(M_S, selected-options);
M_T := run-intention-structure(M).
procedure update-mental-state(M_S, selected-options)
Get the selected-triggers from the selected-options;
For each selected-trigger in selected-triggers do
     Case type of selected-trigger is
          BEL-ADD, BEL-REM, DES-ADD, DES-REM, INT-ADD, INT-REM:
               process-updates(M_S, selected-options);
          ACT-SUC, ACT-FAL, DES-SUC, DES-FAL, PLN-SUC, PLN-FAL:
               process-results(M_S, selected-options);
Return the updated belief and intention states.
```

Fig. 2. Procedures for reactive reasoner and update-mental-state

value. The interpreter responds to triggers by altering its internal state. The type of a trigger can be any one of the following: BEL-ADD, BEL-REM, DES-ADD, DES-REM, INT-ADD, INT-REM, ACT-SUC, PLN-SUC, ACT-FAL, PLN-FAL. The first six types were discussed elsewhere [15] and they initiate the addition or removal of beliefs, desires, and intentions, respectively. The last four events are generated when a primitive act or plan succeeds or fails, respectively.

Depending on the type of the trigger the value of the trigger can be belief, desire, intention, or plan values. An *option* is a trigger followed by either a plan-name or an action. An *end* is either a desire to achieve/recognize a proposition or to execute/observe a primitive action. The *means* is an instantiated plan which contains the plan's mode, name, invocation, context, effects, and the current-node. An *end tree* consists of an end followed by a list of zero or more means trees. Each means tree is a partial unwinding of a means to satisfy the end. A *means tree* consists of a means followed by a list of one or more end trees. Each end tree is a partial unwinding of the end that needs to be satisfied for the program control to move to the next node. An *intention tree* is a special type of end tree with at least one means tree.

The procedure **reactive-reasoner** is given in Figure 2 and operates as follows. The reasoner is called in each cycle with the set of plans P, the current mental-state M_S, and a list of triggers that contain both internal and external events. The result of running the reactive reasoner is given by the mental-state M_T. The agent first generates all options for all the triggers in the trigger list, i.e., the plans that can satisfy a given trigger. The agent then deliberates on these options and selects a subset of these options to which it commits. The mental state of the agent is then updated with respect to this set of selected options. Finally, the agent runs the updated intention structure resulting in a new mental-state.

procedure process-updates(M_S, selected-options)
Get the selected-triggers from the selected-options;
For each selected-trigger in selected-triggers do
    Case type of selected-trigger is
        BEL-ADD, BEL-REM:
            Update B_S with the proposition given in the selected-trigger
        DES-ADD:
            Create a new intention tree as follows:
                (a) The root of the intention tree is the value of the selected trigger
                (b) The children of the root are the various plans that appear in
                the selected options for the selected trigger
            Add the new intention tree to I_S
        DES-REM:
            Remove the intention tree whose root node matches the value of the
            selected trigger from I_S.
        INT-ADD:
            Add the end as the child for the means in the intention tree.
            Add the plans that appear in the selected options as the children for the end.
        INT-REM:
            Remove the end (means) as the child for the means (end) in the intention tree.
Return the updated belief and intention states.

Fig. 3. Procedure for processing updates

We take the mental-state to be the set of beliefs and intentions. Note that we have not taken the desires to be part of the mental state. This is because desires can be of two types: *intrinsic desires* that arise from within the agent and *secondary desires* that arise as a result of the agent attempting to satisfy its intrinsic desires. Once the agent chooses to satisfy an intrinsic desire it becomes an intention and all the secondary desires are represented as intentions. An intrinsic desire that is not chosen by the agent is dropped and is not maintained from one time point to the other. However, the beliefs and intentions of the agent are maintained from one time point to the next. This implementation is a reflection of the different axioms of commitment we have adopted for desires, as compared to beliefs and intentions [15].

The procedure **option-generator** is straightforward and generates the set of applicable plans for each trigger. The **deliberate** procedure could include domain-independent decision-theoretic analysis for choosing options with the highest expected utility or domain-dependent heuristics for filtering the options. Space precludes us from discussing the **deliberate** procedure further.

An important part of the **reactive-reasoner** is the procedure that updates the mental-state of the agent by processing the options selected by the deliberate procedure. This involves a case-by-case analysis of all the events and how they change the mental-state of the agent. The details of the procedure **update-mental-state** is also given in Figure 2.

The procedure **update-mental-state** processes updates (such as belief, desire, and intention, additions and removals) and results (such as the success or failure of actions, desires, and plans). These are given by the procedures **process-updates** and **process-results**. The current mental-state of the agent is then returned by the **update-mental-state** procedure.

The events that change the set of beliefs of the agent are assumed to be internally consistent and therefore we avoid the problem of belief revision. Adding a desire results in the agent creating a new intention tree with the desire as the end and the plans that can satisfy that end as the means of the intention tree. Removing a desire (which should be an end at the root of some existing intention tree) results in the entire intention tree being removed from the intention structure. Adding or removing intentions results in changes to the existing intention tree. The addition and removal of beliefs, desires, and intentions are given by the procedure **process-updates** in Figure 3.

Processing the success or failure of actions, desires, and plans is given by the procedure **process-results** in Figure 4. When a primitive action or desire is successful, either in the execution or recognition modes, the end corresponding to that action or desire is removed from the intention tree. If the parent of this end (i.e., the plan) is currently at an **OR-node** we can remove all the other edges and proceed to the next node, because semantically it is sufficient for one edge to succeed for the **OR-node** to succeed. In the case of an **AND-node** one can proceed to the next node only if the end that succeeded was the last edge of the node. If one proceeds to the next node and the next node is an **END-node** then the entire plan is successful and a successful plan event is posted.

When a primitive action or desire fails in the execution mode, the end corresponding to that action or desire is removed from the intention tree. If the parent of this end (i.e., the plan) is currently at an **AND-node** we can remove all the other edges and fail the plan, because semantically an **AND-node** fails if one of its edges fails. In the case of an **OR-node** one fails the plan only if the end that succeeded was the last edge of the node.

Note that we have not said anything in the case of a primitive action or desire failing in the recognition mode. In other words, the agent is committed to observing/recognizing the primitive actions and desires and will not fail the plan until it has recognized them. In essence, the intentions are in a suspended state waiting for the successful observation of primitive actions to take place. To avoid *blind commitment* towards recognition one can have a time-out on how long the agent waits before abandoning its desires.

At the end of the **update-mental-state** procedure we could have a number of intention trees in the intention structure whose leaves are means (i.e., plans) or ends (i.e., desires or primitive actions). The procedure **run-intention-structure** runs each intention tree in this intention structure. If the leaf node is a primitive action then the agent either executes the action or observes the action. As discussed earlier, in the case of execution the agent performs the action once and the success or failure of the act is informed to the agent by the environment as **ACT-SUC** or **ACT-FAL** events. In the case of recognition if the agent

procedure process-results(M_S, selected-options)
Get the selected-triggers from the selected-options;
For each selected-trigger in selected-triggers do
    Case type of selected-trigger is
        ACT-SUC, DES-SUC:
            Remove the end given by the first argument of the selected trigger and all its children.
            Go to the parent of end which will necessarily be a means.
            Case current node of means is
                OR: Remove all the children of means from the intention tree
                    Make the current node point to the next node in the plan
               AND: If there are no more children of means then make the current
                    node point to the next node in the plan
            If the current node is an END node then
                send PLN-SUC event with means, parent of means, and the intention
                structure as its three arguments
        ACT-FAL, DES-FAL:
            If mode of the selected trigger is EXEC then
                Remove the end given by the first argument of the selected trigger.
                Go to the parent of end which will necessarily be a means.
                    Case current node of means is
                        AND: Remove all the children of means from the intention tree
                             send PLN-FAL event with means, parent of means, and the intention
                           structure as its three arguments
                        OR: If there are no more children of means then
                           send PLN-FAL event with means, parent of means, and the intention
                         structure as its three arguments
        PLN-SUC:
            send BEL-ADD events for all the effects of the plan
            send INT-REM event with the same three arguments as the PLN-SUC event
            send DES-SUC event with the second argument (end), parent of the second
            argument (means), and the third argument (intention-structure) of the PLN-SUC event
        PLN-FAL:
            send INT-REM event with the same three arguments as the PLN-FAL event
            send DES-FAL event with the second argument (end), parent of the second
            argument (means), and the third argument (intention-structure) of the PLN-FAL event
Return the updated belief and intention states.

Fig. 4. Procedure for processing results

cannot observe the successful performance of the action, it waits indefinitely (or until the intention is timed out).

If the leaf node is a plan the agent executes the plan by adding new INT-ADD events for each outgoing edge of the current node. For example, an AND-node which has multiple outgoing edges will have multiple INT-ADD events being generated. These events will then be processed in the next cycle by the **update-mental-state** procedure. A leaf node which happens to be a desire is not processed by the **run-intention-structure** because the `INT-ADD` events for such cases would have been generated by the **update-mental-state** procedure.

The above interpreter allows the executing agent to interleave plans for different independent desires and for the recognizing agent to recognize all these desires, if it wanted to recognize them. However, the recognizing agent cannot recognize past actions of the executing agent. If required this can be handled by having an observation history of fixed length which the recognizing agent can use to "observe" past events.

As the interpreter has the means-end tree of the recognition process, it can infer the beliefs, desires, and intentions of the other agents as well. It is important to note that the agent can recognize the mental-state as the environment is changing. The agent may not always be certain about the mental-state of the other agent (i.e., have a disjunction of mental attitudes), but will nevertheless have useful information about the other agent to direct its own execution.

4 Example

Now we illustrate the means-end reasoning process for reactive plan execution and recognition using an example [9] of making two pasta dishes (See Figure 6). The syntax of the plans is a simplified form of plan specifications in the Procedural Reasoning System (PRS) [4, 7]. We consider a simple example that does not make use of all the features of the algorithms discussed so far. In particular, we do not consider `AND-nodes`, complex context conditions with the mental-state of other agents, and combined recognition and execution within the same plan.

Consider the case where John is making a pasta dish and Mary is observing John making a pasta dish. We assume that both John and Mary have identical plans for making pasta and both are in the kitchen. We start with John having the desire to make a pasta dish and Mary having the desire to recognize the making of a pasta dish. Ideally, we would like Mary to recognize what type of pasta is being made before John has finished making the pasta dish. Figure 5 shows the means-end tree for Mary before any recognition has taken place.

Mary's desire is to recognize the making of a pasta dish by John. This corresponds to the following trigger in the trigger list: (DES-ADD RECOG Mary (! John (made-pasta-dish))). The **reactive-reasoner** procedure first produces the various plans that can respond to this event, namely **Recog. Make Pasta Dish 1** and **Make Pasta Dish 2**. Assuming that the **deliberator** selects these options the **update-mental-state** procedure will process the trigger and create an intention tree whose root node is the desire (i.e., `END1` node of Figure 5) and whose children are the two plans that can satisfy the desire (i.e., `MEANS1` and `MEANS2` nodes of Figure 7 [1]). At this stage the leaves of the intention tree are `MEANS1` and `MEANS2`, both of which are plans. The **run-intention-structure** results in the following events being sent as triggers: (INT-ADD DESIRE RECOG

---

[1] Due to lack of space we have not reproduced all the different fields of the nodes and also have abbreviated DESIRE, PLAN, PRM-ACT, and RECOG to DES, PLN, ACT, and R, respectively.

Mary (! John (made-ordinary-pasta))) and (INT-ADD DESIRE RECOG Mary (! John (made -spaghetti -marinara))). This completes one cycle of the BDI interpreter.

In the next cycle there will be at least two triggers in the trigger list, i.e., the INT-ADD events discussed above. In addition, the trigger list may contain other external events. Options will be generated for all the triggers in the trigger list. The plans Recog. Make Ordinary Pasta and Recog. Make Spaghetti Marinara will respond to the two triggers. The deliberator then selects a subset of these options according to a pre-defined criterion. For example, if the deliberator selects options with the highest priority, then depending on whether the external events have a higher priority or not the deliberator procedure will choose the two INT-ADD events. Assume that there are no other higher priority options. The two INT-ADD events will be processed by the **update-mental-state** procedure. This will result in the intention tree having MEANS3 and MEANS4 as its leaves. The **run-intention-structure** will result in the following two INT-ADD events being sent: (INT-ADD DESIRE RECOG Mary (! John (made-noodles))) and (INT-ADD PRM-ACT RECOG Mary (John (make-spaghetti))). This completes the second cycle of the BDI interpreter.

In the third cycle after the update-mental-state procedure the intention tree will have MEANS5, MEANS6, and END5 as its leaves (see Figure 5). As END5 is a primitive action the agent can observe the act. If the act has just occurred, an ACT-SUC will be triggered. If it has not occurred the intention to observe make spaghetti will be suspended, waiting for the event to occur. Let us assume that the latter holds and the intention tree has MEANS5, MEANS6, and END5 as its leaves.

In the fourth cycle after the update-mental-state procedure the intention tree will have END6, END7, and END5 as its leaves (see Figure 5). In other words, the agent is waiting to observe (John make-fettucini) and (John make-spaghetti). Assume that Mary observes John making fettucini. This will result in a ACT-SUC event being generated for END6.

In the next cycle the ACT-SUC event will be processed; node END6 will be removed and the current node of the Recog. Make Fettucini plan will be updated. However, as the current node is an END-node of the plan a PLN-SUC event will be generated for MEANS5. This will result in a DES-SUC event being generated for END4. Processing this will result in both its children, MEANS5 and MEANS6, being removed, as the desire has been achieved. The current node of MEANS3 will be updated. This process continues till the entire plan has been recognized.

5 Comparison and Conclusions

Plan Recognition: Agent-oriented programming, where users write reactive plans to guide an agent through the planning process has received a great deal of attention in recent years. The Procedural Reasoning System [4] is one of the oldest systems that allows one to write agent programs with respect to the beliefs, desires, and intentions of an agent. There are two primary differences between the

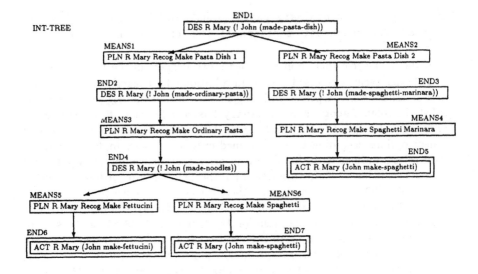

Fig. 5. Means-End Tree before the recognition of *make-fettucini*

BDI interpreter given above and the Procedural Reasoning System [4]. First, we have presented a propositional variant of the PRS interpreter which is a restricted first-order system with full unification. Second, PRS is a reactive planning system that expands the means-end tree in a depth-first manner for efficiency reasons. As a result the interpreter makes use of a means-end stack to represent a top-level intention and all its children. Therefore in PRS, reactive recognition has to be explicitly programmed by the user.

Early work by Allen and Perrault [1] and more recently by Litman and Allen [10] treat plan recognition as the reverse process of planning (in the classical sense). Kautz [9] presents a formal approach to plan recognition which makes use of an *event hierarchy* to guide the recognition process. However, the above approaches are not amenable to reactive recognition and are not embedded within the mental-state of an agent.

The work that comes closest to our approach in this respect is the work of Tambe and Rosenbloom [18]. He describes an approach, called RESC (Real-time Situated Commitments), that allows real-time tracking of the mental-state of other agents. His approach relies on making a commitment to one of the means-end paths and then proceeding with the recognition. In that sense, the approach turns out to be a particular metal-level deliberation strategy to prune the search space.

More recently, Huber *et. al.* [6] have shown how to take reactive plans written in PRS and convert them into a probabilistic belief network. This network is then used to recognize the mental state of other agents. The approach presented in this paper is similar, but allows the user to write recognition plans, rather than deriving them from the reactive plans.

Applications: The initial motivation for developing a theory of reactive recognition was the need to infer the mental-states of opponents in an air-combat modelling system. The application of the above algorithms to this domain is discussed elsewhere [16]. In this domain one has to infer not only the positional information of enemy aircraft but also recognize their mission goals and the tactics used to achieve these goals. Mental state recognition is far more important in such adversarial reasoning scenarios than in collaborative reasoning, as agents cannot communicate and elicit further information from opponents. However, reactive recognition can be used fruitfully in a number of collaborative domains such as discourse understanding [10] and automated tutoring systems.

The algorithms discussed in this paper have been implemented in Gofer (GOod For Equational Reasoning), a functional programming environment for a small Haskell-like language [8]. After further extensions and experimentation with respect to a first-order version, the algorithms will be included in the production version of the Distributed Multi-Agent Reasoning System (dMARS), a successor to PRS, written in C++.

In summary, the primary contribution of this paper is in presenting an integrated approach to reactive plan execution and recognition that is applicable to a broad class of problems where the environment is changing rapidly and the agents are resource-bounded. By embedding these processes within a framework of the mental-state of agents, they also facilitate mental-state recognition. The novelty of the approach is in the use of plans as recipes and as mental attitudes to guide and constrain the reasoning processes of agents.

Acknowledgements I would like to thank Mike Georgeff for valuable comments on earlier drafts of this paper and the the anonymous reviewers for their useful suggestions. This research was supported by the Cooperative Research Centre for Intelligent Decision Systems under the Australian Government's Cooperative Research Centres Program.

References

1. J. F. Allen and C. R. Perrault. Analyzing intention in utterances. *Artificial Intelligence*, 15:143–178, 1980.
2. M. E. Bratman, D. Israel, and M. E. Pollack. Plans and resource-bounded practical reasoning. *Computational Intelligence*, 4:349–355, 1988.
3. P. R. Cohen and H. J. Levesque. Intention is choice with commitment. *Artificial Intelligence*, 42(3), 1990.

4. M. P. Georgeff and A. L. Lansky. Procedural knowledge. In *Proceedings of the IEEE Special Issue on Knowledge Representation*, volume 74, 1986.

5. P. Haddawy and S. Hanks. Issues in decision-theoretic planning: Symbolic goals and numeric utilities. In *Proceedings of the DARPA Workshop on Innovative Approaches to Planning, Scheduling, and Control*, 1990.

6. M. J. Huber, E. H. Durfee, and M. P. Wellman. The automated mapping of plans for plan recognition. In *Proceedings of the Tenth Conference on Uncertainty in Artificial Intelligence (UAI-94)*, pages 344–351. Morgan Kaufmann Publishers, San Mateo, CA, 1994.

7. F. F. Ingrand, M. P. Georgeff, and A. S. Rao. An architecture for real-time reasoning and system control. *IEEE Expert*, 7(6), 1992.

8. M. P. Jones. The implementation of the gofer functional programming system. Technical Report Research Report, YALE/DCS/RR-1030, Department of Computer Science, Yale University, New Haven, CT 8285, USA, 1994.

9. H. Kautz. A circumscriptive theory of plan recognition. In P. R. Cohen, J. Morgan, and M. E. Pollack, editors, *Intentions in Communication*. MIT Press, Cambridge, MA, 1990.

10. D. J. Litman and J. Allen. Discourse processing and commonsense plans. In P. R. Cohen, J. Morgan, and M. E. Pollack, editors, *Intentions in Communication*. MIT Press, Cambridge, MA, 1990.

11. M. E. Pollack. Plans as complex mental attitudes. In P. R. Cohen, J. Morgan, and M. E. Pollack, editors, *Intentions in Communication*. MIT Press, Cambridge, MA, 1990.

12. A. S. Rao. Means-end plan recognition: Towards a theory of reactive recognition. In *Proceedings of the Fourth International Conference on Principles of Knowledge Representation and Reasoning (KRR-94)*, Bonn, Germany, 1994.

13. A. S. Rao and M. P. Georgeff. Asymmetry thesis and side-effect problems in linear time and branching time intention logics. In *Proceedings of the Twelfth International Joint Conference on Artificial Intelligence (IJCAI-91)*, Sydney, Australia, 1991.

14. A. S. Rao and M. P. Georgeff. Modeling rational agents within a BDI-architecture. In J. Allen, R. Fikes, and E. Sandewall, editors, *Proceedings of the Second International Conference on Principles of Knowledge Representation and Reasoning*. Morgan Kaufmann Publishers, San Mateo, CA, 1991.

15. A. S. Rao and M. P. Georgeff. An abstract architecture for rational agents. In C. Rich, W. Swartout, and B. Nebel, editors, *Proceedings of the Third International Conference on Principles of Knowledge Representation and Reasoning*. Morgan Kaufmann Publishers, San Mateo, CA, 1992.

16. A. S. Rao and G Murray. Multi-agent mental state recognition and its application to air-combat modelling. In *Proceedings of the 13th International Distributed Artificial Intelligence Workshop (DAI-13)*. AAAI Press, 1994.

17. S. Russell and S. Subramanian. On provably ralphs. In E. Baum, editor, *Computational Learning and Cognition: The Third NEC Research Symposium*. SIAM Press, 1992.

18. M. Tambe and P. S. Rosenbloom. Resc: An approach for real-time, dynamic agent tracking. In *Proceedings of the Fourteenth International Joint Conference on Artificial Intelligence (IJCAI-95)*, Montreal, Canada, 1995.

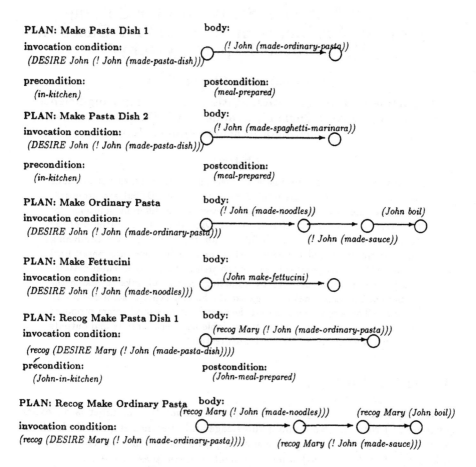

PLAN: Make Pasta Dish 1
invocation condition:
(DESIRE John (! John (made-pasta-dish)))

precondition:
(in-kitchen)

body:
(! John (made-ordinary-pasta))

postcondition:
(meal-prepared)

PLAN: Make Pasta Dish 2
invocation condition:
(DESIRE John (! John (made-pasta-dish)))

precondition:
(in-kitchen)

body:
(! John (made-spaghetti-marinara))

postcondition:
(meal-prepared)

PLAN: Make Ordinary Pasta
invocation condition:
(DESIRE John (! John (made-ordinary-pasta)))

body:
(! John (made-noodles))
(! John (made-sauce))
(John boil)

PLAN: Make Fettucini
invocation condition:
(DESIRE John (! John (made-noodles)))

body:
(John make-fettucini)

PLAN: Recog Make Pasta Dish 1
invocation condition:
(recog (DESIRE Mary (! John (made-pasta-dish))))

precondition:
(John-in-kitchen)

body:
(recog Mary (! John (made-ordinary-pasta)))

postcondition:
(John-meal-prepared)

PLAN: Recog Make Ordinary Pasta
invocation condition:
(recog (DESIRE Mary (! John (made-ordinary-pasta))))

body:
(recog Mary (! John (made-noodles)))
(recog Mary (! John (made-sauce)))
(recog Mary (John boil))

Fig. 6. Plan library for making pastas

Neural Network Strategies for Solving Synthesis Problems in Non-conflict Cases in Distributed Expert Systems*

Minjie Zhang and Chengqi Zhang

Department of Mathematics, Statistics and Computing Science
University of New England, Armidale, N.S.W. 2351, Australia
{minjie,chengqi }@neumann.une.edu.au

Abstract

In this paper, two neural network mechanisms for synthesis of solutions in non-conflict cases in distributed expert systems (DESs) are proposed. The ideas are: inputs of the neural network are different solutions for the same problem from different expert systems in DESs; outputs of the neural network are the final solutions for the problem after combining different solutions which should be the same as the human experts' final solutions. The first point is to set up the architecture of the neural network and train the neural network by adjusting weights of links to match the outputs of the neural network against the human experts' solutions for all patterns. The second point is that the neural network mechanism proposed in this paper can accommodate the variable number of inputs and outputs without changing neural network architecture.

1 Introduction

In a distributed expert system (DES), each cooperative member is an expert system (ES) with its own domain knowledge and inference engine. Each ES can work individually for solving some specific problems and can also cooperate with other expert systems (ESs) when they deal with complex problems.

Due to the limitation of knowledge, the ability of problem solving in single ESs, and the uncertain features of problems, some complex tasks need to be solved by diverse expertises in order to increase reliability of the solution. A typical example is when "several doctors diagnose the same patient".

If a subtask is allocated to more than one ES, each ES could obtain a solution. For example, two expert systems (ESs) predict a potential earthquake in a particular area. ES_1 believes that the possibility of the potential earthquake being class 4 is $x_1 = 0.8$, while ES_2 believes that the possibility of the potential earthquake being class 4 is $x_2 = 0.6$. It is not good practice for a DES to give a user diverse solutions to the same subtask. The problem here is how to obtain the final uncertainty if more than one uncertainty for the same solution exists. The synthesis strategies are responsible for synthesizing the uncertainties of the solution from different ESs to produce the final uncertainty of the solution.

Consider the following two cases based on the above example.

---

*This research was supported by a large grant from the Australian Research Council (A49530850).

Case (1) Two ESs obtain the uncertainties of the solution (a class 4 earthquake in a particular area) $x_1 = 0.8$ and $x_2 = 0.6$, respectively, based on the same geochemical results. This case demonstrates a belief conflict between ES_1 and ES_2 because they obtained the same solution with different uncertainties given the same evidences [8]. The final uncertainty $S(x_1, x_2)$ should be between x_1 and x_2, i .e. $\min(x_1, x_2) \leq S(x_1, x_2) \leq \max(x_1, x_2)$ where S is a synthesis strategy [8].

Case (2) ES_1 predicts a class 4 earthquake in an area with uncertainty of $y_1 = 0.8$ based on the evidence from geophysical experiment and ES_2 obtains the same solution with uncertainty of $y_2 = 0.6$ based on the geological structure of this area. The final uncertainty for the solution of a class 4 earthquake in this case should be bigger than any one of y_1 and y_2, i.e. $S(y_1, y_2) \geq max\{y_1, y_2\}$, because two ESs obtain the same solution from different evidences, so this solution is more reliable than if it comes from the same evidence [8].

From the above two cases, we can see that in different synthesis cases, we should use different strategies.

Roughly speaking, synthesis cases can be divided into two parts, conflict cases and non-conflict cases.

Traditional methods of solving belief conflicts in DESs are synthesis strategies which include: uncertainty management strategy developed by Khan [2]; a synthesis strategy for heterogeneous DESs introduced by Zhang [6]; and an improved synthesis strategy which considers both uncertainties and authorities of ESs proposed by Liu [3]. Based on the above research, a new comprehensive synthesis strategy was proposed to overcome several limitations of above strategies [7].

We have also identified a number of potential synthesis cases [8] in DESs and have proposed two strategies to solve belief conflicts in DESs [7]. Then, we have proposed necessary conditions of synthesis strategies and synthesis strategies for non-conflict cases [10].

These traditional methods of synthesis strategies are based on the mathematical analysis and the proposer of the strategy must know the detail mathematical relationships between the inputs (different solutions from different ESs) and the outputs (final solutions).

Generally, a good synthesis strategy is one which can almost always get the same final solutions as the human experts do in the real world. The synthesis strategies based on mathematical analysis do not work well in some cases since the relationship between the inputs and outputs are too hard to be summarized. The reason is that human experts do not strictly follow the mathematical formula and they sometimes use their experiences to solve synthesis problem.

However, a neural network is a right mechanism (after training) to simulate the relationship between the inputs and the outputs if reasonable numbers of inputs and corresponding outputs are known. Recently, we proposed a dynamic neural network synthesis strategy solving belief conflicts in DESs [9]. To our knowledge, this neural network strategy is the first attempt at using neural network mechanism for solving synthesis problems in DESs. For the cases with enough patterns, neural network strategy is successful. We now extend this idea to non-conflict synthesis cases.

This paper describes two neural network strategies working in two different non-conflict synthesis cases and their performances. The paper is organized as follows. In Section 2, basic problems and different synthesis cases are described. In Section 3, a neural network synthesis strategy for overlap synthesis cases is proposed. In Section 4, a neural network synthesis strategy for disjoint synthesis cases is discussed. Finally, in Section 5, this research is concluded.

2   Basic problems and different synthesis cases

2.1   Problem description

Let's see an example first. Suppose there are three ESs (e.g. ES_1, ES_2, ES_3) to decide the identity of the organism for a specific patient. ES_1 says that it is pseudomonas with uncertainty 0.36 and proteus with uncertainty -0.9, ES_2 says that it is pseudomonas with uncertainty 0.5 and serratia with uncertainty 0.4, and ES_3 says that it is serratia with uncertainty 0.1 and proteus with uncertainty 0.85. Because ES_1 doesn't mention serratia, we believe that ES_1 has no idea about it. We can represent this unknown by using uncertainty 0 in the EMYCIN model [5]. Then the above solutions are represented in Table 1.

	Pseudomonas	Serratia	Proteus
ES_1	0.36	0	-0.9
ES_2	0.5	0.4	0
ES_3	0	0.1	0.85

Table 1: The uncertainties for each attribute value obtained by the ESs.

The purpose of the synthesis of solutions here is how to decide the final uncertainty of pseudomonas, serratia, and proteus according to Table 1.

We now formally describe the problems. Suppose there are n ESs in a DES to evaluate the values of an attribute of an object (e.g. in a medical DES, the identity of an organism infecting a specific patient). The solution for an ES_i can be represented as

$$(< object >< attribute > (V_1 \ CF_{i1} \ A_i) \ (V_2 \ CF_{i2} \ A_i) \ ... \ (V_m \ CF_{im} \ A_i))$$
2.1

where V_j $(1 \leq j \leq m)$ represents jth possible value, CF_{ij} $(1 \leq i \leq n, 1 \leq j \leq m)$ represents the uncertainty for jth value from ES_i, A_i represents the authority of ES_i, and m indicates that there are m possible values for this attribute of the object. For example, there are 6 possible values for the face-up of a dice.

From the synthesis point of view, all ESs are concerned with the same attribute of an object. So we will only keep the attribute values, uncertainties, and authorities in the representation. Here is the representation of m possible values with uncertainties from n ESs.

$$\begin{bmatrix} (V_1\ CF_{11}\ A_1)(V_2\ CF_{12}\ A_1)...(V_m\ CF_{1m}\ A_1) \\ (V_1\ CF_{21}\ A_2)(V_2\ CF_{22}\ A_2)...(V_m\ CF_{2m}\ A_2) \\ .... \\ (V_1\ CF_{n1}\ A_n)(V_2\ CF_{n2}\ A_n)...(V_m\ CF_{nm}\ A_n) \end{bmatrix} \qquad 2.2$$

The synthesis strategy is responsible for obtaining final uncertainties of the vector $(V_1\ CF_{*1}\ A_*)(V_2\ CF_{*2}\ A_*)\ ...\ (V_m\ CF_{*m}\ A_*)$ based on matrix 2.2 where * indicates the synthesis result from corresponding values with the subscriptions of $1, 2, ..., n$ in the same place such as $CF_{*m} = S(CF_{1m}, A_1, CF_{2m}, A_2, ..., CF_{nm}, A_m)$ where S is a synthesis function.

Let's see another example. Suppose that there are 4 ESs using the PROBABILITY inexact reasoning model [1] to decide the identity of the organism infecting a specific patient. Each ES obtains six values for the organism. The uncertainties of values obtained by ESs are shown in Table 2.

	$Value_1$	$Value_2$	$Value_3$	$Value_4$	$Value_5$	$Value_6$	Authorities
ES_1	0.05	0.60	0.20	0.05	0.05	0.05	0.95
ES_2	0.01	0.65	0.04	0.03	0.17	0.03	0.85
ES_3	0.03	0.67	0.03	0.05	0.10	0.02	0.70
ES_4	0.06	0.40	0.14	0.19	0.18	0.02	0.80

Table 2: The uncertainties in the PROBABILITY model

The aim of the synthesis strategies here is to decide the final uncertainty of $Value_1$ to $Value_6$ from Table 2.

2.2 Potential synthesis cases in DESs

From the introduction, we know, for the same uncertainties, the synthesis results can be different. That is, when we develop synthesis strategies, the first problem is to identify the different synthesis cases in DESs [8].

There are four potential synthesis cases in DESs based on the relationship between evidence sets of a solution from different ESs [8] which are: (a) conflict synthesis cases, (b) inclusion synthesis cases, (c) overlap synthesis cases, and (d) disjoint synthesis cases. We call (b), (c), and (d) as non-conflict synthesis cases.

Here we briefly describe the definitions of four synthesis cases.

(a) A *conflict synthesis case* [8] occurs when the original evidence sets of a proposition from different ESs are equivalent, but different ESs produce the same solution with different uncertainties. That is, for a proposition B, if there exist $E_{B_i} = E_{B_j}$, where E_{B_i} is the original evidence set to deduce the proposition B in ES_i, E_{B_j} is the original evidence set to deduce the proposition B in ES_j, and $CF_i \neq CF_j$, where CF_i is the uncertainty of the proposition B from ES_i, CF_j is the uncertainty of the proposition B from ES_j.

(b) An *inclusion synthesis case* occurs when the original evidence set of a proposition from one ES is a subset of the original evidence set of another ES.

Formally, for a proposition B, $E_{B_i} \subset E_{B_j}$ or vice versa where E_{B_i} in ES_i and E_{B_j} in ES_j.

(c) A *partial overlap synthesis case* occurs when the original evidence sets of a proposition from different ESs are not equivalent, but the intersection of original evidence sets is not empty. Formally, for a proposition B, $E_{B_i} \cap E_{B_j} \neq \phi$, $E_{B_i} \cap E_{B_j} \neq E_{B_i}$, and $E_{B_i} \cap E_{B_j} \neq E_{B_j}$ where E_{B_i} in ES_i and E_{B_j} in ES_j.

(d) A *disjoint synthesis case* occurs when the intersection of original evidence sets of a proposition from different ESs is empty. Formally, for a proposition B, $E_{B_i} \cap E_{B_j} = \phi$, where E_{B_i} in ES_i, and E_{B_j} in ES_j.

So far, most papers have been concerned with a conflict synthesis case. However, solving non-conflict cases is as important as solving a conflict case.

In this paper, we will introduce two neural network strategies working in overlap and disjoint synthesis cases, respectively.

3 A neural network synthesis strategy for overlap synthesis cases

In this section, a neural network synthesis strategy to solve synthesis problems in overlap synthesis cases will be introduced.

3.1 Basic principles

Generally speaking, there are two methods to build neural networks: to train neural networks by learning patterns, or to train neural networks by learning rules. Since it is difficult to find rules for synthesis of solutions in DESs (recall Section 1), we use learning patterns to create our neural network.

The ideas are: the inputs of the neural network are the matrix of uncertainty values of the same attribute of an object from different ESs (recall matrix 2.2 or Table 2); the outputs of neural network should be a vector of uncertainty values of the attribute of the object which are the final uncertainty values of the DESs after synthesizing input uncertainties. The closer the vector of uncertainty values to human experts' final uncertainty values, the better. We would like to investigate whether a neural network can be trained to solve synthesis problems in overlap synthesis cases if enough patterns are given. We should collect enough number of patterns (input matrix and corresponding output vector) from human experts; to set up a neural network architecture (input layer nodes, hidden layer nodes, and output layer nodes), to decide an activation function, and to train this neural network by adjusting the weights of links to accommodate all patterns.

The following activation function is used.

$$a_j(t+1) = \frac{1}{1 + e^{\left(\sum_i w_{ij} o_i(t) - \theta_j\right)}}$$

where $a_j(t)$ is the activation of unit j in step t, $o_i(t)$ is the output of unit i in step t, w_{ij} is the weight of link from unit i to unit j, and θ_j is the threshold of unit j.

One of the important research issues in neural networks is how to adjust the weight of links to get the desired system behavior [4].

In our strategy, we use a backpropagation learning algorithm to train the network since it is most commonly used for classification problem solving.

The backpropagation weight update rule is described as follows:

$$\Delta w_{ij} = \eta \delta_j o_i$$

$$\delta_j = \begin{cases} f'_j(net_j)(t_j - o_j) & \text{if unit } j \text{ is a output-unit} \\ f'_j(net_j) \sum_k \delta_k w_{jk} & \text{if unit } j \text{ is a hidden-unit} \end{cases}$$

where η is a learning factor, δ_j is the error of unit j, t_j is the teaching input of unit j, o_i is the output of preceding unit j, i is the index of a predecessor to the current unit j with link w_{ij} from i to j, j is an index of the current unit and k is an index of a success or to the current unit j with link w_{jk} from j to k.

3.2 A dynamic neural network mechanism

A neural network should have $n * m$ input nodes and m output nodes (recall matrix 2.2 for the meaning of n and m) in order to solve our problem. For a DES, the number of ESs (n) is relatively stable. However, the possible number of values (m) for each attribute varies from case to case.

It is not practical to train a different neural network every time when the number of values is changed. In this subsection we will propose a method to build a fixed neural network architecture for synthesis of solutions in overlap synthesis cases to adapt the problem with variable numbers of inputs and outputs.

Suppose k is the maximum number of values of all attributes in a DES. The idea is to set up a neural network with $n * k$ input nodes and k output nodes. The training patterns can be $n * 2$ inputs with 2 outputs, $n * 3$ inputs with 3 outputs, ..., and $n * k$ inputs with k outputs. If the number of values for some attributes is less than k, the default values "0" are chosen to substitute these absent values as inputs and corresponding outputs. If the neural network after training can accommodate all (enough number of) patterns, we believe that this neural network can work well for untested cases. We have built a neural network to test the above ideas. We choose $n = 4$, $k = 6$. The input layer has $4 * 6$ nodes and output layer has 6 nodes.

There are three layers in this network. The input layer is a 4*6 matrix, the hidden layer is a 2*10 matrix, and the output layer is a 1*6 vector. The neural network is fully connected. The input of the neural network is the matrix of uncertainty values of the same attribute of an object from 4 ESs (recall matrix 2.2); and the output of the neural network is a vector of uncertainty values of attribute of the object which are the final uncertainty values of DESs after synthesizing input uncertainties in a overlap synthesis case.

The architecture of the neural network is shown in Figure 1.

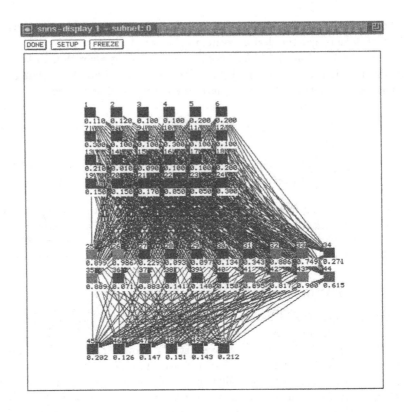

Figure 1: The architecture of a neural network for overlap synthesis cases

Table 3 and Table 4 are examples of training patterns with different default values.

Input	$Value_1$	$Value_2$	$Value_3$	$Value_4$	$Value_5$	$Value_6$
ES_1	0.11	0.12	0.10	0.10	0.20	0.20
ES_2	0.30	0.10	0.10	0.30	0.10	0.10
ES_3	0.21	0.01	0.09	0.10	0.10	0.20
ES_4	0.15	0.15	0.17	0.05	0.05	0.03
Output	0.162	0.162	0.184	0.133	0.133	0.227

Table 3: An example of training patterns without default values

3.3 Training and testing results

So far we haven't collected real data from human experts. What we would like to investigate in this paper is that whether a neural network can converge for a set of complicated artificial data. If it succeeds, it would be a positive signal to encourage us to collect real data from human experts to train neural networks.

Input	$Value_1$	$Value_2$	$Value_3$	$Value_4$	$Value_5$	$Value_6$
ES_1	0	0	0.20	0.20	0.30	0.30
ES_2	0	0	0.30	0.30	0.20	0.20
ES_3	0	0	0.05	0.16	0.09	0.66
ES_4	0	0	0.18	0.28	0.25	0.25
Output	0	0	0.151	0.134	0.161	0.554

Table 4: An example of training patterns with two default values

In this paper, there are 102 different patterns created by the mathematical synthesis strategy for overlap synthesis cases [10] considering different possibilities. 92 of them are used as training patterns and other 10 untrained patterns are used as test patterns.

The training results are shown in Table 5.

Training Cycles	SSE	SSE/o-units
10,000	0.12175	0.02029
20,500	0.02695	0.00449
64,000	0.01617	0.00234
100,000	0.00907	0.00151

Table 5: Training results

The error value δ_{pq}^0 for the qth output neuron acting of input vector (or matrix) p is calculated as follows:

$$\delta_{pq}^0 = (y_{pq} - o_{pq})$$

where y_{pq} is the desired output of the qth unit of vector (or matrix) p and o_{pq} is the final output of the qth unit of vector (or matrix) p. The total error of the network E_p^0 is described by the summation of the error values for each output unit k:

$$E_p^0 = \frac{1}{2}\sum_{k=1}^m (\delta_{pq}^0)^2 = \frac{1}{2}\sum_{k=1}^m (y_{pq} - o_{pq})^2$$

where m is the number of nodes in output layer.

In Table 5, SSE is the sum-squared error of the learning function for all nodes and SSE/o-units is the result of SSE divided by the number of output units. The training result is good. When we select 10 patterns from 92 training patterns to test this network, the mean error is less than 1.5%. For 10 untrained patterns, the mean error is on 2.8% level.

The percentage of the error from test patterns in each stage and both maximum and mean errors are shown in Table 6 and Table 7.

Figure 2 shows the distribution of errors from test patterns.

range of error	percentage
0 – 0.01	15/60
0.01 – 0.02	12/60
0.02 – 0.03	8/60
0.03 – 0.04	9/60
0.04 – 0.05	5/60
0.05 – 0.06	5/60
0.06 – 0.07	2/60
0.07 – 0.08	2/60
0.08 – 0.09	1/60
0.09 – 0.11	1/60

Table 6: The percentage of error in each stage

Maximum error	0.111
Mean error	0.028
Number of data	60

Table 7: Information of final errors

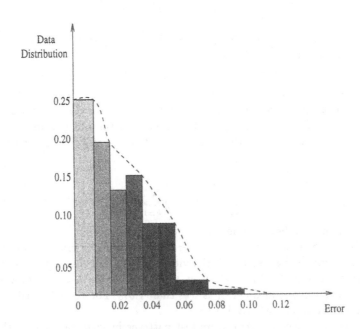

Figure 2: The distribution of the test errors

Table 6 shows that 82% $(15/60 + 12/60 + 8/60 + 9/60 + 5/60 = 49/60 = 0.82)$ of errors from output data are below 0.05, 93.3% $(15/60 + 12/60 + 8/60 + 9/60 + 5/60 + 5/60 + 2/60 = 0.933)$ of errors from output data are below 0.07, and only 6.7% of errors from output data are between 0.07 to 0.11. This result can verify that the neural network mechanism can also use in overlap synthesis cases.

Table 8 shows the synthesis result from both mathematical synthesis strategy [10] and this neural network strategy.

Input (Uncertainties from ESs)	V_1	V_2	V_3	V_4	V_5	V_6
ES_1	0.05	0.60	0.20	0.05	0.05	0.05
ES_2	0.01	0.65	0.04	0.03	0.17	0.03
ES_3	0.03	0.67	0.03	0.05	0.10	0.02
ES_4	0.06	0.40	0.14	0.19	0.18	0.02
Final result (Mathematical strategy)	0.035	0.555	0.122	0.098	0.139	0.051
Output (Neural network strategy)	0.040	0.600	0.099	0.089	0.130	0.060

Table 8: The comparison of the final results between two strategies

Now let us compare the output of neural network strategy with the final result of the mathematical synthesis strategy in overlap synthesis cases (see Table 8). The mean error of the neural network strategy is 0.014 and maximum error is 0.045. That means the synthesis result from neural network can be trusted.

4 A neural network synthesis strategy for disjoint synthesis cases

In this section, we will propose a neural network synthesis strategy for disjoint synthesis cases.

4.1 Architecture of a neural network

We also use the idea of dynamic mechanism to develop this strategy (recall Section 3.2).

There are also three layers in the neural network which are input layer, hidden layer, and output layer. Both the input layer and the output layer are the same sizes as the neural network for overlap synthesis cases. The hidden layer is a 7*7 matrix. Please see Subsection 3.1 for the meaning of the input and the output of a neural network synthesis strategy. The neural network is also fully connected.

Figure 3 is the architecture of the neural network.

The learning algorithm is also backpropagation (recall Section 3.1). There are 95 patterns created by the mathematical synthesis strategy for disjoint synthesis cases [10] to be used as training patterns. There are 7 untrained patterns used as test patterns.

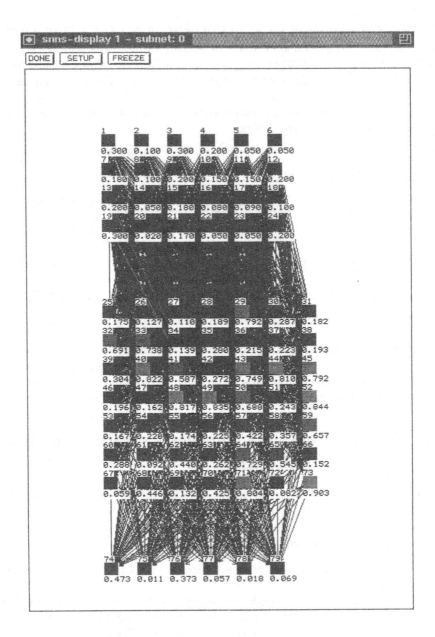

Figure 3: The architecture of a neural network for disjoint synthesis cases

4.2 Result analyses

The neural network was trained by 53,000 cycles. Table 9 is the training result.

Training Cycles	SSE	SSE/o-units
10,000	0.11413	0.01902
20,000	0.02629	0.00438
53,000	0.01040	0.00173

Table 9: Training results

Please see Subsection 3.3 for the meaning of SSE and SSE/o-units.

When we randomly selected 10 patterns from training patterns to test the network, the mean error is around 1%. For 7 untrained patterns, the percentage of the error from testing results in each stage is shown in Table 10. Both maximum and mean errors are shown in Table 11.

range of error	percentage
0 – 0.01	11/42
0.01 – 0.02	7/42
0.02 – 0.03	4/42
0.03 – 0.04	3/42
0.04 – 0.05	3/42
0.05 – 0.06	4/42
0.06 – 0.07	2/42
0.07 – 0.08	1/42
0.08 – 0.09	1/42
0.09 – 0.11	1/42

Table 10: The percentage of error in each stage

Maximum error	0.107
Mean error	0.036
Number of data	42

Table 11: Information of final errors

Figure 4 figures out the distribution of test errors from test patterns.

From Table 11, we can see that the mean error for test patterns is on 3.6% level. From Table 10, we can find that 67% of errors are below 0.05, 92.9% of errors are below 0.08, and 7% of errors are from 0.08–0.107.

We use the same example as in Section 3.3 to synthesize a disjoint synthesis case by this neural network strategy. Table 12 shows the synthesis result by using this neural network strategy.

Table 13 shows results of final uncertainties between the mathematical synthesis strategy and the neural network strategy for the same disjoint synthesis case.

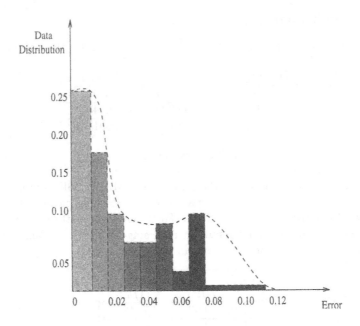

Figure 4: The distribution of the test errors

Input	$Value_1$	$Value_2$	$Value_3$	$Value_4$	$Value_5$	$Value_6$
ES_1	0.05	0.60	0.20	0.05	0.05	0.05
ES_2	0.01	0.65	0.04	0.03	0.17	0.03
ES_3	0.03	0.67	0.03	0.05	0.10	0.02
ES_4	0.06	0.40	0.14	0.19	0.18	0.02
Output	0.001	0.941	0.033	0.006	0.002	0.000

Table 12: Synthesis result by the neural network strategy

Strategies	$Value_1$	$Value_2$	$Value_3$	$Value_4$	$Value_5$	$Value_6$
Mathematical model	0.002	0.913	0.026	0.010	0.047	0.002
Neural network	0.001	0.941	0.033	0.006	0.002	0.000

Table 13: The comparison of the final results between mathematical and neural network strategies

The above table shows that the maximum error from neural network strategy is 4.5% and the mean error from neural network strategy is only 1.4%. This result indicates that a neural network synthesis strategy is a good way to solve synthesis problems if enough training patterns are available.

Two neural network strategies introduced in this paper were implemented by the Stuttgart Neural Network Simulator [4].

5   Conclusion

In this paper, we introduced two neural network mechanisms to solve synthesis problems in overlap and disjoint synthesis cases and tested that a fixed neural network architecture can solve the problems with a variable number of inputs and outputs of neural networks.

This is the first time a neural network mechanism has been used to solve synthesis problems in non-conflict cases in DESs. The contributions are as follows.

(1) The neural network is easy to build because we need to know enough patterns of inputs and outputs only without knowing the relationship between them.

(2) The results of synthesis of solutions in non-conflict cases by using neural network mechanism are good so far because trained neural network can simulate patterns very well.

(3) A fixed neural network architecture can be used to solve the problem with a variable number of inputs and outputs. This result makes it practical to use the neural network mechanism to solve synthesis problems in DESs, even though the number of inputs and outputs of the neural network is changeable. This result can also be used in other application fields of neural networks.

(4) If the neural network converges and there are enough patterns available, the neural network strategy is better than mathematical formula because it can simulate human expertise quite closely.

Further work will include two directions: (1) when the data in the real world are collected and used to train neural networks, we will see whether the performance is still good enough; (2) when the number of patterns increase extremely, we will see whether neural networks still converge in reasonable time.

References

[1] R. Duda, P. Hart and N. Nilsson (1976). Subjective Bayesian Method for Rule-based Inference System, *AFIPS*, **45**, pp. 1075-1082.

[2] N. A. Khan and R. Jain (1985). Uncertainty Management in a Distributed Knowledge Based System. In *Proceedings of 5th International Joint Conference on Artificial Intelligence*, Los Angeles, pp. 318-320.

[3] D. Liu, F. Zheng, Z. Ma and Q. Shi (1992). Conflict Resolution in Multi-ES cooperation Systems. *Journal of Computer Science and Technology*, Vol. 7, No. 2, pp. 144-152.

[4] N. Mache, R. Hubner, G. Mamier, and M. Vogt (1993). *SNNS*, Report No. 3/93, Institute for Parallel and distributed High Performance Systems, Germany.

[5] W. Van Melle (1980). A Domain-Independent System That Aids in Constructing Knowledge-Based Consultation Programs, *Ph.D. Dissertation, Report STAN-CS-80-820*, Computer Science Department, Stanford University, CA.

[6] C. Zhang (1992). Cooperation under Uncertainty in Distributed Expert Systems, *Artificial Intelligence*, **56**, pp. 21-69.

[7] M. Zhang and C. Zhang (1994). A Comprehensive Strategy for Conflict Resolution in Distributed Expert Systems, *Australian Journal of Intelligent Information Processing Systems*, **Vol. 1**, No. 2, pp. 21-28.

[8] M. Zhang and C. Zhang (1994). Synthesis of Solutions in Distributed Expert Systems. In *Artificial Intelligence - Sowing the Seeds for the Future*, edited by Chengqi Zhang, John Debenham and Dickson Lukose, World Scientific Publishers, Singapore, pp. 362-369.

[9] M. Zhang and C. Zhang (1995). A Dynamic Neural Network Mechanism for Solving Belief Conflict in Distributed Expert Systems. In *Proceedings of Sixth Australian Conference on Neural Networks*, Sydney, Australia, pp. 105-108.

[10] M. Zhang and C. Zhang (1995). The Synthesis Strategies for Non-conflict Cases in Distributed Expert Systems. *AI'95*, World Scientific Publishers, Singapore, pp. 195-202.

A Linearly Quasi-Anticipatory Autonomous Agent Architecture: Some Preliminary Experiments

Paul Davidsson

Dept. of Computer Science, Lund University, Box 118, S–221 00 Lund, Sweden
E-mail: paul.davidsson@dna.lth.se

Abstract. This report presents some initial results from simulations of a linear quasi-anticipatory autonomous agent architecture (ALQAAA), which correspond to a special case of a previously suggested general architecture of anticipatory agents. It integrates low-level reaction with high-level deliberation by embedding an ordinary reactive system based on situation-action rules, called the Reactor, in an anticipatory agent forming a layered hybrid architecture. By treating all agents in the domain (itself included) as reactive agents, this approach drastically reduces the amount of search needed while at the same time requiring only a small amount of heuristic domain knowledge. Instead it relies on a linear anticipation mechanism, carried out by the Anticipator, to achieve complex behaviours. The Anticipator uses a world model (in which the agent is represented only by the Reactor) to make a sequence of one-step predictions. After each step it checks whether the simulated Reactor has reached an undesired state. If this is the case it will modify the actual Reactor in order to avoid this state in the future. Results from both single- and multi-agent simulations indicate that the behaviour of ALQAAA agents is superior to that of the corresponding reactive agents. Some promising results on cooperation and coordination of teams of agents are also presented. In particular, the linear anticipation mechanism is successfully used for conflict detection.

1 Introduction

In this report we will present some initial results from experiments on a linearly quasi-anticipatory autonomous agent architecture. This architecture corresponds to a special case of the general architecture of anticipatory agents suggested by Astor, Davidsson, and Ekdahl [1, 4].

1.1 Hybrid Agent Architectures

In the last couple of years it has become widely acknowledged that an intelligent autonomous agent must have the capability of both high-level deliberation and low-level reaction. This insight can be seen as a synthesis of two earlier rival lines of thought: the first argued that agent cognition should be based on sophisticated deliberation, typically planning, and the second that it should be based only on primitive reactive behaviour specified for instance by situation-action rules. As it turns out, the weaknesses of reactive approach correspond closely to the strengths of deliberative approach and vice versa, i.e., reactive agents are fast but "dumb" (and do not need an explicit world model) whereas deliberative agents are "smart" but slow (and need a detailed world model).

Moreover, a combined, or *hybrid*, approach seems to model human functioning closer than the purely reactive approach, which resembles that of more primitive animals.

As Hanks and Firby [8] point out, two categories of hybrid agent architectures can be distinguished; *uniform* architectures employ a single representation and control scheme for both reaction and deliberation, whereas *layered* architectures use different representations and algorithms implemented in separate layers to perform these functions. Most uniform architectures, for example PRS [10], do not make any specific commitments on how reaction and deliberation should be interleaved. In addition, Ferguson [5] argues that "There are a number of other reasons for advocating a layered control approach, including increased behavioural robustness and operational concurrency, as well as improved program comprehensibility and system testability and analysability." (p.48)

However, as pointed out by Wooldridge and Jennings [17] there is a weakness in the existing suggestions for hybrid architectures. They argue that:

> One potential difficulty with such architectures, however, is that they tend to be *ad hoc* in that while their structures are well-motivated from a design point of view, it is not clear that they are motivated by any deep theory. (p.26)

In order to improve this situation, a layered hybrid approach based on the theory of anticipatory systems will be presented in the next section.

1.2 Anticipatory Agents

According to Rosen [15], an anticipatory system is " ... a system containing a predictive model of itself and/or of its environment, which allows it to change state at an instant in accord with the model's predictions pertaining to a latter instant." (p.339) Thus, such a system uses the knowledge concerning future states to decide which actions to take in the present.

Let us describe a simple class of anticipatory systems suggested by Rosen [14]. It contains an ordinary causal (i.e., non-anticipatory) dynamic system, S. With S he associates another dynamical system, M, which is a model of S. It is required that the sequence of states of M are parameterized by a time variable that goes faster than real time. That is, if M and S are started out at some time t_0, then after an arbitrary time interval Δt, M's sequence of states will have proceeded $t_0 + \Delta t$. In this way, the behaviour of M predicts the behaviour of S: by looking at the state of M at time t, we get information about the state that S will be in at some later time than t. In addition, M is equipped with a set E of effectors which allows it to operate either on S itself, or on the environmental inputs to S, in such a way as to change the dynamical properties of S. If S is modified the effectors must also update M. This class of anticipatory systems is illustrated in Fig. 1. Rosen argues that this would be an anticipatory system in the strict sense if M were a perfect model of S (and if the environment were constant or periodic). However, as M in general is not a perfect model of S, he calls the behaviour of such a system *quasi-anticipatory*.

Then, how should these predictions be used to modify the properties of S? Rosen [14] argues that this could be done in many ways, but suggests that the following is the simplest:

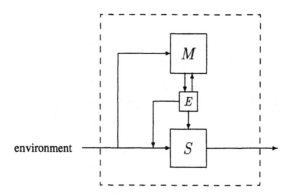

Fig. 1. The basic architecture of a class of anticipatory systems.

Let us imagine the state space of S (and hence of M) to be partitioned into regions corresponding to "desirable" and "undesirable" states. As long as the trajectory in M remains in a "desirable" region, no action is taken by M through the effectors E. As soon as the M-trajectory moves into an "undesirable" region (and hence, by inference, we may expect the S-trajectory to move into the corresponding region at some later time, calculable from a knowledge of how the M- and S-trajectories are parameterized) the effector system is activated to change the dynamics of S in such a way as to keep the S-trajectory out of the "undesirable" region. (p.248)

However, to transfer the concept of anticipatory systems into an agent framework there are some additions and changes that we have found necessary. First, we need a meta-level component that runs and monitors the model, and evaluates the predictions made to decide how to change S or the input to S.[1] Thus, we also include the effectors, E, in this meta-level component that we here will call the *Anticipator*. Second, in order to predict future environmental inputs to S we need to extend the model M to include also the environment. This inclusion is in line with later work of Rosen (cf. [15]). In sum, an *anticipatory agent* consists mainly of three entities: an object system (S), a world model (M), and a meta-level component (Anticipator). The object system is an ordinary (i.e., non-anticipatory) dynamic system. M is a description of the environment *including* S, but excluding the Anticipator. The Anticipator should be able to make predictions using M and to use these predictions to change the dynamic properties of S. Although the different parts of an anticipatory agent certainly are causal systems, the agent taken as a whole will nevertheless behave in an anticipatory fashion.

When implementing an anticipatory agent, what should the three different components correspond to, and what demands should be made upon these components? To begin with, it seems natural that S should correspond to some kind of reactive system similar to the ones mentioned above. We will therefore refer to this component as the *Reac-*

---

[1] We will in the following regard both these types of changes as changes to S.

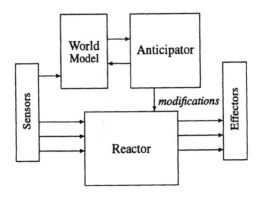

Fig. 2. The basic architecture of an anticipatory agent.

tor. It should be a fast system in the sense that it should be able to handle routine tasks instinctively and, moreover, it should have an architecture that is both easy to model and to change. The Anticipator would then correspond to a more deliberative meta-level component that is able to "run" the world model faster than real time. When doing this it must be able to reason about the current situation compared to the predicted situations and its goals in order to decide whether (and how) to change the Reactor.

The resulting architecture is illustrated in Fig. 2. To summarize: The sensors receive input from the environment. This data is then used in two different ways: (1) to update the World Model and (2) to serve as stimuli for the Reactor. The Reactor reacts to these stimuli and provides a response that is forwarded to the effectors, which then carry out the desired action(s) in the environment.[2] Moreover, the Anticipator uses the World Model to make predictions and on the basis of these predictions the Anticipator decides if, and what, changes of the dynamical properties of the Reactor are necessary. Every time the Reactor is modified, the Anticipator should, of course, also update the World Model accordingly. Thus, the working of an anticipatory agent can be viewed as two concurrent processes, one reactive at the object-level and one more deliberative at the meta-level.

2   A Basic Class of Anticipatory Agents

In this section some initial results from experiments with A Linearly Quasi-Anticipatory Agent Architecture (ALQAAA) will be presented. ALQAAA corresponds to a special case of the architecture described in the last section. In particular, it follows Rosen's suggestion of the simplest way of deciding when to change the Reactor, i.e., by dividing the state space into desired and undesired regions.

---

[2] Do not confuse these effectors with those discussed above that perform actions that modifies the object system S.

Some simple ALQAAA-agents and a testbed has been implemented.[3] The problem domain has deliberately been made as simple as possible in order to make the principles of anticipatory behaviour as explicit as possible.

2.1 Agent Implementation

The Reactor and the Anticipator are run (asynchronously) as two separate processes. The Reactor process is given a high priority whereas the Anticipator is a low priority process that runs whenever the Reactor is "waiting" (e.g., for an action to be performed). Since the Reactor is able to preempt the Anticipator at any time, reactivity is always guaranteed. Thus, the Anticipator has to be a kind of *anytime algorithm* [2], or rather anytime process, in that it should always be able to return a result when it is interrupted.[4] The appropriateness of using anytime algorithms in autonomous agent contexts where real-time requirements are common has been pointed out by, for example, Zilberstein and Russell [18] and Bresina and Drummond [3].

The Reactor carries out a never ending cycle of: perception of the environment (i.e., the situation), action selection by situation-action (stimuli-response) rules, and performance of action. Rather than having explicit goals, the Reactor's goals are implicitly represented in its collection of situation-action rules. The basic algorithm of the Reactor is given below:

> **procedure** REACTOR;
> **while** true **do**
> Percepts ← Percieve;
> Action ← SelectAction(Percepts);
> Perform(Action);

The Anticipator, on the other hand, carries out a never ending cycle of anticipation sequences. Each such sequence begins with making a copy of the World Model, which as mentioned earlier is a description of the environment containing the agent as a physical entity in the environment, and a copy of the agent's current set of reaction rules. These are then used to make a sequence of one-step predictions. After each prediction step, it is checked whether the simulated agent has reached an undesired state, or whether it has achieved the goal. If it has reached an undesired state, the Reactor will be manipulated in order to avoid reaching this state. Thus, this functioning corresponds to that of the simplest kind of anticipatory system suggested by Rosen. The basic algorithm of the Anticipator is as follows:

---

[3] They have been implemented in the object-oriented language Simula on Sun SparcStation running Solaris 2.3. Local class packages were used to achieve concurrency (Simlib IOProcesses) and graphical interface to X (WindowPackage).

[4] According to Dean and Boddy [2], the main characteristics of anytime algorithms are that "... (i) they lend themselves to preemptive scheduling techniques (i.e., they can be suspended and resumed with negligible overhead), (ii) they can be terminated at any time and will return some answer, and (iii) the answers returned improve in some well-behaved manner as a function of time." (p.52)

```
procedure ANTICIPATOR;
while true do
    WorldModelCopy ← WorldModel;
    ReactorCopy ← Reactor;
    UndesiredState ← false;
    while not UndesiredState and not GoalAchieved(WorldModelCopy) do
        Percepts ← WorldModelCopy.Percieve;
        Action ← ReactorCopy.SelectAction(Percepts);
        WorldModelCopy.Perform(Action);
        UndesiredState ← Evaluate(WorldModelCopy);
    if UndesiredState then
        Manipulate(Reactor);
```

Note that since the behaviour of the Reactor in each situation is determined by situation-action rules, the Anticipator always "knows" which action the Reactor would have performed. Also the environment (including all other agents) is treated as being purely reactive. Thus, since everything is governed by situation-action rules, the anticipation mechanism requires no search, or in other words, the anticipation is *linear*. It should also be noted that the goal of the agent is not limited to have only a singular goal. In a multi-goal scenario, some of the changes (manipulations) of the Reactor should only hold for a limited interval of time (e.g., until the current goal has been achieved). Otherwise, there is a danger that these changes might prevent the agent to achieve other goals.

In more formal terms a linearly quasi-anticipatory agent can be specified as a tuple:

$$\langle \mathcal{R}, \mathcal{W}, \mathcal{U}, \mathcal{M} \rangle$$

where

> $\mathcal{R}$ is the set of situation-action rules defining the Reactor.
> $\mathcal{W}$ is the description of the environment (the world model).
> $\mathcal{U}$ is the set of undesired states.
> $\mathcal{M}$ is the set of rules describing how to modify $\mathcal{R}$.

The Anticipator is defined by $\mathcal{U}$ and $\mathcal{M}$. For each element in $\mathcal{U}$ there should be a corresponding rule in $\mathcal{M}$, which should be applied when an undesired state of this kind is anticipated. Thus, we need in fact also a function, $f : U \rightarrow M$, that determines which rule for modifying the Reactor that should be applied given a particular type of undesired state. However, as this function typically is obvious from the specification of $\mathcal{U}$ and $\mathcal{M}$, it will not be described explicitly. Moreover, in all simulations described below, $\mathcal{W}$ will consist of the positions of all obstacles, targets, and agents present in the environment.

Using these terms, the function Evaluate can be described as checking whether the current anticipated state belongs to $\mathcal{U}$, and Manipulate as first applying f on the anticipated undesired state and then using the resulting rule from $\mathcal{M}$ to modify $\mathcal{R}$.

2.2 The Testbed

The agent's environment is a two-dimensional grid (10×10) in which a number of unit-sized square obstacles forms a maze. In addition to these static objects, there are two kinds of dynamic objects: agents, which can move about in the maze, and targets, which can be removed by an agent.

The goal of an agent is to pick up the target(s). To be able to pick up a target, the agent must be in the same position as the target. The agent is able to move in four directions (north, south, east, and west), unless there is an obstacle that blocks the way. The agent is always able to perceive the direction to the target, and whether there are any obstacles immediately north, south, east, or west of the agent.[5]

Almost identical environments have been used in other experiments, for instance by Sutton [16]. It is also similar to the Tileworld [13] (if we interpret the targets as holes) except for that: (i) following Kinny and Georgeff [11] our testbed has no tiles as they only make the simulations unnecessarily complex and (ii) some randomness have been excluded (e.g., the pattern-less appearance and disappearance of holes) since it makes much of the deliberation pointless (i.e., prediction becomes impossible, cf. Hanks [9]). In Section 2.4 the environment is generalized into a multi-agent scenario (as suggested by, for instance, Zlotkin and Rosenschein [19]). The advantages and disadvantages of this kind of testbeds have been discussed at length by Hanks, Pollack and Cohen [9].

2.3 Single Agent Simulations

In this section some experiments on single agents will be presented. We will compare the performance of reactive agents with that of ALQAAA agents containing such agents as their Reactor.

Let us consider a simple reactive agent that has only one simple situation-action rule: *move to the free position closest to the target*. In other words, it first tries to move to that adjacent position yielding the greatest reduction in distance to the target. If there is an obstacle in this position, it tries the direction of the remaining three that reduces the distance if there are one, else it tries the position that increases the distance the least and so on. A reactive agent of this kind will behave rather well in many not too complicated mazes. However, as illustrated in Fig. 3 (a) there are also very simple mazes in which it behaves poorly. Starting out in the position marked A, it will move two positions towards the target (T) and then "loop" between two positions as marked in the figure.

Let us now consider an ALQAAA agent that has this reactive agent as its Reactor. That is, $\mathcal{R} = \{move\ to\ the\ free\ position\ closest\ to\ the\ target\}$. We define the undesired states as those in which the agent is trapped in a loop, and if such a state is detected by the Anticipator, the Reactor is manipulated in such a way that it will not enter this state from now on. Thus, we have that: $\mathcal{U} = \{being\ in\ a\ loop\}$ and $\mathcal{M} = \{avoid\ the\ position\ in\ the\ loop\ closest\ to\ the\ target\}$.[6] An ALQAAA agent of this kind will then behave as showed in Fig. 3 (b). The Anticipator will by anticipation detect the loop before entering it and

---

[5] Think of the obstacles as being made of glass: it is possible to see the targets through the obstacles but not the obstacles themselves, which only can be perceived by tactile sensors.

[6] Which position in the loop to avoid can, of course, be selected according to other principles.

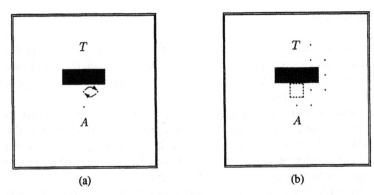

Fig. 3. The behaviour of (a) the Reactive agent and (b) the ALQAAA agent. A indicates the agent's initial position and T the position of the target.

make the Reactor avoid the position closest to the target (marked by a dashed box). This is sufficient for the Reactor to find its way to the target.

We have compared the performance of a reactive and an ALQAAA agent of the kinds described above in a series of experiments. There was one target in the environment and the number of obstacles varied between 0 and 35. In each experiment the positions of the agent, the target and the obstacles were randomly chosen. In order to avoid too many trivial scenarios there was also a requirement that the distance between the agent and the target should be at least five unit-lengths. Moreover, only scenarios in which it is possible for the agent to reach the target were selected. From the result in Fig. 4, we see that the more complex the environment gets, the more useful is the anticipatory behaviour. If there are no obstacles at all, even the Reactor will, of course, always find the target.

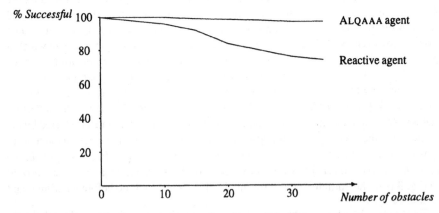

Fig. 4. Comparison between Reactive and ALQAAA agents. (200 runs of each multiple of five)

This ALQAAA agent is able to reach the target (when this is possible) in almost all kinds of mazes. However, as Fig. 4 shows, there are some in which it will not succeed and they are typically of the kind depicted in Fig. 5 (a). The reactive agent will in this

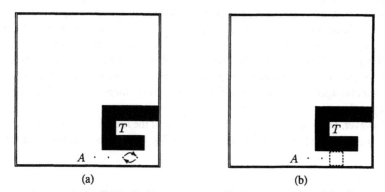

(a)   (b)

Fig. 5. The behaviour of (a) Reactive agent and (b) ALQAAA agent.

case be trapped in a loop whereas the ALQAAA agent will detect this loop beforehand and block the position as shown in Fig. 5 (b). This implies that the only possible way to the target is blocked and the agent will never reach the target.

The problem with this Anticipator is that it is too eager to block a position. The reason for this is that the Reactor is inclined to "turn around" as soon as it is not decreasing the distance to the target. If we just augment the Reactor's situation-action rule with the condition that it should only change its direction 180° when there are no other alternatives (i.e., if there are obstacles in the three other directions), we will get a reactive agent that solves this problem. If we take this Reactor and the same Anticipator as used in the last example, we get an ALQAAA agent which seems always to reach the target (if the Anticipator is given enough time to anticipate, that is). This Reactor (i.e., $\mathcal{R} = \{move\ to\ the\ free\ position\ closest\ to\ the\ target,\ but\ do\ not\ turn\ around\ if\ not\ forced\ to\}$) will be used in all the experiments that follows.

2.4 Multi-Agent Simulations

In this section some experiments in multi-agent domains are presented. We will point out the advantages of being anticipatory, both when competing and when cooperating with other agents. Although the experiments have been carried out with two competing or cooperating agents in order to make things as clear as possible, it would be trivial to extend the experiments to permit a larger number of agents.

Competing Agents The main idea here is that an ALQAAA agent should use its knowledge about the behaviour of other agents in order to detect future situations in which other agents interfere with the agent's own intentions (i.e., goals). If such a situation is detected, the Anticipator should manipulate the Reactor in order to minimize the probability that this situation will ever occur.

We will evaluate this approach in the same testbed as above but with two agents and more than one target in the environment. Agent A should be regarded as "our" agent, whereas agent B represents the agent with which it competes. The goal of both the agents is to pick up as many targets as possible. In addition to the basic algorithm described above, the Anticipator needs a model of the agent B which it uses to predict B's actions in the same manner as it predicts its own (i.e., A's) actions. When the Anticipator realizes that B will reach a target before A, it notifies the Reactor that it should ignore this target. Thus, we have that: $\mathcal{U} = \{$ *being in a loop, pursuing targets that presumably will be picked up by another agent*$\}$ and $\mathcal{M} = \{$ *avoid the position in the loop closest to the target, avoid the target that presumably will be picked up by another agent*$\}$.

However, let us start with two reactive agents of the kind described above. An environment containing three targets is described in Fig. 6 (a). If the agents start at the same

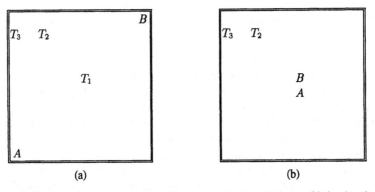

Fig. 6. The behaviour of two competing reactive agents: (a) the initial state (b) the situation after 8 time steps, agent B has picked up target T_1.

time the following will happen. Both agents perceive that T_1 is their closest target and they both head towards it. As B is somewhat closer to T_1 than A, it will reach it first and pick it up (see Fig. 6 (b)). B will then head for T_2 which now is the closest target. A will also head for T_2 following B. B then reaches T_2, picks it up, and heads for the last target T_3 with A still behind. Eventually B will pick up also T_3. Thus, B gets all the targets and A gets none.

If we, on the other hand, let A be an ALQAAA agent and start in the same position as above, it will soon detect that B will be the first to reach T_1. So, A will avoid this target and instead head towards T_3 (which is the next closest target to A). It will reach T_3 at the same time as B reaches T_1. When the agents have picked up their targets, there is only one target left (T_2). Since A is closest to T_2, it will reach it first. Thus, by anticipating the behaviour of both itself and the other agent, A will in this case pick up two targets whereas B only picks up one.

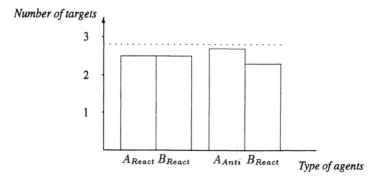

Number of targets

3

2

1

$A_{React}\ B_{React}$ $A_{Anti}\ B_{React}$ *Type of agents*

Fig. 7. Comparison between two sets of competing agents. To the left are both A and B reactive agents and to the right is A an ALQAAA agent and B a reactive agent. The vertical axis indicates the number of targets picked up by an agent (averages over 1000 runs). The optimal number of targets that A is able to pick up in this situation (i.e., given B's behaviour) is illustrated by the dashed line.

There are also some quantitative results on the superiority of ALQAAA agents when competing with reactive agents. In the experiments there were 30 obstacles, 5 targets and two agents. In the first session both the agents were reactive and in the second there was one ALQAAA agent competing with a reactive one. The results shown in Fig. 7 tell us that the performance indeed is improved (being almost optimal) when the agent behaves in an anticipatory fashion.

Cooperating Agents We shall now see how ALQAAA agents can be used for cooperation. The task for the two agents is to pick up all the targets in shortest possible time. It does not matter which agent that picks up a particular target.

To begin with, we apply the agents in the last example (i.e., A is an ALQAAA agent and B a reactive agent) to the situation described in Fig. 8 (a). As these agents are not cooperating, their global behaviour will (as one might expect) not be optimal. What will happen is that both agents initially head towards the same target (T_1). When agent A reaches T_1 we have the situation depicted in Fig. 8 (a). The other targets will then be approached in the same fashion, with one agent following the other. As a result, it will take these non-cooperating agents 15 time-steps to pick up all the targets.

Cooperating agents, on the other hand, should be able to make use of the fact that the ALQAAA agent knows that it will pick up the two closest targets. One way of doing this is to let agent A send a message to agent B (which still is a reactive agent) when it believes that it will pick up a particular target. This message contains the information that agent B should avoid this target. Thus, we add *"other agent pursuing target that presumably will be picked up by me"* to $\mathcal{U}$ and *"send message to other agent that it should avoid the target"* to $\mathcal{M}$.

When this method is applied to the previous example, agent A will detect that it will pick up targets T_1 and T_2 and sends therefore messages to B that these should be avoided. B then directly heads towards T_3, which is the only remaining target that A

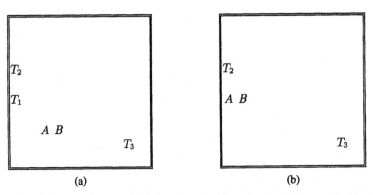

Fig. 8. The behaviour of two non-cooperating agents one ALQAAA (A) and one reactive (B). (a) the initial state (b) the situation after 4 time steps, agent A has picked up target T_1.

will not reach before B. As a result, this system of cooperating agents will use only 6 time-steps to pick up all the targets.

The quantitative results (using 30 obstacles, 5 targets and two agents) are summarized in Fig. 9. We see that when the two agents are cooperating they come close to optimal behaviour.

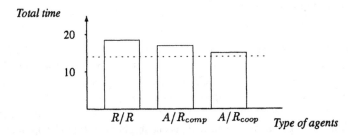

Fig. 9. Comparison between three sets of agents in terms of the total time it takes to collect all the five targets (averages over 1000 runs). R/R denotes two reactive agents, A/R_{comp}, one reactive and one ALQAAA agent competing, and A/R_{coop}, one reactive and one ALQAAA agent that are cooperating. The optimal time for two agents to collect all targets in this situation is illustrated by the dashed line.

In the scenario described here, it is only agent A that is an ALQAAA agent whereas B is an ordinary reactive agent. Even if we also let B be an ALQAAA agent with the same model of the world, we would not increase the performance. The reason for this is that both agents would have made the same predictions and therefore send messages to each other about things that both agents have concluded by themselves. However, in a more realistic setting where the agents do not have exactly the same information about the world, such an approach would probably be fruitful. In such a case things get quite complicated if one ALQAAA agent simulates the anticipatory behaviour of another ALQAAA agent which in turn simulates first agent's anticipatory behaviour. The solution we suggest is to simulate only the reactive component of other agents and when an agent modifies its reactive component, it should communicate (e.g., broadcast) information about this modification to the other agents. In this way we are still able to make linear anticipations. This approach can be contrasted with the Recursive Modeling Method suggested by Gmytrasiewicz and Durfee [7] in which an agent modeling another agent includes that agent's models of other agents and so on, resulting in a recursive nesting of models.

As we have seen the behaviour of the ALQAAA agents implemented has not been optimal. However, we have deliberately chosen very simple Reactors and Anticipators for the purpose of illustrating how easily the performance can be improved by embedding a Reactor in an ALQAAA agent. It should be clear that it is possible to develop more elaborate $\mathcal{R}$, $\mathcal{U}$, and $\mathcal{M}$ components that produce behaviour closer to the optimal.

3 Discussion

We have shown the viability of an approach for designing autonomous agents based on the concept of anticipatory systems called ALQAAA in a simple navigation task. Adaptation to the environment is performed by letting the Anticipator component of the agent manipulate the Reactor component according to anticipated future states.

Compared to traditional planning, anticipation as described here (there may be other ways to anticipate) is a more passive way of reasoning. The ALQAAA agent just tries to predict what will happen if nothing unexpected occurs, whereas a planning agent actively evaluates what will happen when a number of different actions are performed. The result is that planning agents will rely heavily on search, whereas ALQAAA agents will not. The main reason for this is that all agents in the environment (also the ALQAAA agent itself) are treated as being reactive.

In addition, it is interesting to note the small amount of heuristic domain knowledge that is given to the Reactor and the Anticipator (i.e., $\mathcal{R}$, $\mathcal{U}$, and $\mathcal{M}$). Thus, this approach drastically reduces the amount of search needed while at the same time requiring only a small amount of heuristic domain knowledge. Instead, it relies on a linear anticipation mechanism to achieve a more complex behaviour.

3.1 Related Work

The main task of the Anticipator is to avoid undesired states whereas the main task of the Reactor is to reach the desired state(s). In other words, the Anticipator's goals are goals of maintenance and prevention rather than of achievement. Compare this to Minsky's suppressor-agents, discussed within his Society of Mind framework [12], which waits until a "bad idea" is suggested and then prevents the execution of the corresponding action. However, there is a big difference, suppressor-agents are not predictive. The Anticipator takes actions beforehand so that the bad idea never will be suggested! Thus, an Anticipator can be regarded as predictive suppressor-agent.

In conformity with the Sequencer component in Gat's ATLANTIS architecture [6], the Anticipator can be viewed as being based on the notion of *cognizant failures* (i.e., a failure that can be detected by the agent itself). However, an Anticipator detects these failures in a simulated reality (i.e., a model of the world), whereas a Sequencer has to deal with real failures.

The notions of Reactor and Anticipator have some similarities with the Reactor and Projector components in the ERE architecture suggested by Bresina and Drummond [3]. In particular, the Reactor in ERE is able to produce reactive behaviour in the environment independently, but also takes advise from the Projector based on the Projector's explorations of possible futures. However, the Projector is more similar to a traditional planner in that it is based on search through a space of possible Reactor actions (a third component, the Reductor, is introduced to constrain this search), whereas the Anticipator simulates the behaviour of the Reactor in its environment linearly (i.e., without search). Moreover, the Anticipator's main task is to avoid undesired states, whereas the Projector in the ERE tries to achieve desired states.

There are also some similarities to the DYNA architecture [16] if we let the Reactor correspond to DYNA's Policy component and the Anticipator to its Evaluation function. In DYNA two kinds of rewards can be identified: *external* rewards, which are those that the Evaluation function gets from the environment (this kind of rewards is not required by an ALQAAA agent), and *internal* rewards, which are those that the Policy gets from the Evaluation function (these can be compared with the manipulations that the Anticipator performs on the Reactor). However, there are many disadvantages with the DYNA architecture compared to ALQAAA agents: (i) The planning process in DYNA requires search (in fact, random search) when it internally tests the outcome of different actions. Moreover, several trials are typically necessary whereas ALQAAA agents only need one. (ii) Even if the goal is changed only slightly (e.g., the target is moved one position), the learning in DYNA must start again from scratch. (iii) It is not clear how to implement the Evaluation function. In the initial experiments (cf. [16]), it was implemented using tables where each possible state is represented. This approach is clearly not viable in realistic environments. A more appropriate approach is the one used in ALQAAA agents where only *categories* of undesired states have to be defined, which is often much easier than to define complex reward functions.

3.2 Limitations of Experiments

The problems that the ALQAAA agents solved above can certainly be solved by other methods, but the point to be made is that we can qualitatively enhance the abilities of a reactive agent by embedding it in an ALQAAA agent. However, there are several obvious limitations to the application presented in this paper: (i) the environment is quite static (the only events that take place not caused by the agent itself are those caused by other agents), (ii) the agents have perfect models of the world,[7] (iii) the agents have perfect sensors and the outcome of an action is deterministic, and (iv) this is just a simulation

---

[7] Thus, the behaviour of the agents in the experiments could be regarded as anticipatory rather than quasi-anticipatory.

(the agent is neither embodied nor situated) and thus escaping the hard problems of perception and uncertainty. Moreover, only a single domain has been investigated. Future work includes evaluation of the approach in other domains to see in which types of applications it performs well and whether there are any in which it is not appropriate.

References

1. P. Davidsson, E. Astor, and B. Ekdahl. A framework for autonomous agents based on the concept of anticipatory systems. In *Cybernetics and Systems '94*, pages 1427–1434. World Scientific, 1994.
2. T. Dean and M. Boddy. An analysis of time-dependent planning. In *AAAI-88*, pages 49–54, 1988.
3. M. Drummond and J. Bresina. Anytime synthetic projection: Maximizing the probability of goal satisfaction. In *AAAI-90*, pages 138–144, 1990.
4. B. Ekdahl, E. Astor, and P. Davidsson. Towards anticipatory agents. In M. Wooldridge and N.R. Jennings, editors, *Intelligent Agents — Theories, Architectures, and Languages, Lecture Notes in Artificial Intelligence 890*, pages 191–202. Springer Verlag, 1995.
5. I.A. Ferguson. *TouringMachines: An Architecture for Dynamic, Rational, Mobile Agents.* PhD thesis, University of Cambridge, 1992.
6. E. Gat. Integrating planning and reacting in a heterogeneous asynchronous architecture for controlling real-world mobile robots. In *AAAI-92*, pages 809–815, 1992.
7. P.J. Gmytrasiewicz and E.H. Durfee. Rational coordination in multiagent environments through recursive modeling. *(submitted for publication)*, 1995.
8. S. Hanks and R.J. Firby. Issues and architectures for planning and execution. In *DARPA Workshop on Innovative Approaches to Planning, Scheduling and Control*, pages 59–70, San Mateo, CA, 1990. Morgan Kaufmann.
9. S. Hanks, M. Pollack, and P. Cohen. Benchmarks, testbeds, controlled experimentation, and the design of agent architectures. Technical Report 93–06–05, Department of Computer Science and Engineering, University of Washington, 1993.
10. F.F. Ingrand, M.P. Georgeff, and A.S. Rao. An architecture for real-time reasoning and system control. *IEEE Expert*, 7(6):34–44, 1992.
11. D.N. Kinny and M.P. Georgeff. Commitment and effectiveness of situated agents. In *IJCAI-91*, pages 82–88, 1991.
12. M. Minsky. *The Society of Mind.* Simon and Schuster, 1986.
13. M.E. Pollack and M. Ringuette. Introducing the Tileworld: Experimentally evaluating agent architectures. In *AAAI-90*, 1990.
14. R. Rosen. Planning, management, policies and strategies: Four fuzzy concepts. *International Journal of General Systems*, 1:245–252, 1974.
15. R. Rosen. *Anticipatory Systems – Philosophical, Mathematical and Methodological Foundations.* Pergamon Press, 1985.
16. R.S. Sutton. First results with Dyna, an integrated architecture for learning, planning and reacting. In W.T. Miller, R.S. Sutton, and P.J. Werbos, editors, *Neural Networks for Control*, pages 179–189. MIT Press, 1990.
17. M.J. Wooldridge and N.R. Jennings. Intelligent agents: Theory and practice. *Knowledge Engineering Review*, 10(2):115–152, 1995.
18. S. Zilberstein and S.J. Russell. Anytime sensing, planning and action: A practical model for robot control. In *IJCAI-93*, pages 1402–1407, 1993.
19. G. Zlotkin and J.S. Rosenschein. Coalition, cryptography, and stability: Mechanisms for coalition formation in task oriented domain. In *AAAI-94*, pages 432–437, 1994.

Cooperation Among Opportunistic Agents on a Distributed Environment

Orlando Belo[1] and José Neves[1]

Departamento de Informática, Escola de Engenharia, Universidade do Minho
Largo do Paço, 4709 Braga Codex, PORTUGAL
{obelo,jneves}@di-ia.uminho.pt

Abstract. Often agent-based applications, namely those related with diagnosis, control or classification problems, have to deal with unpredictable situations. This is the case where the agents must be able to call for effective problem solving strategies based on previous experiences. Indeed as result of past experiences and interaction, agents develop social behaviour (promoting inter-agent cooperation, mutual help or information exchange). Building on these ideas and going through models of human behaviour the design and development of a distributed multi-agent system environment was accomplished. This paper presents an overall description of the system, with emphasis on the expert agent classes, their architecture and models of interaction and cooperation.

1 Introduction

Different frames of mind may be observed in humans when faced with a problem. An entering upon a problem's scenario, looking for clues that may help in the problem's solution, it is a very common one. Such a strategy may eventually provide the basis to develop an effective plan to attack efficiently a problem and solve it. Reacting to stimulus generated from environment changes is another typical form of human attitudes. This allows for humans to choose, for a given event, the best way to react and consequently to apply a better strategy. The difficulty to establish a comparative model to evaluate properly these two approaches is significative. A verdict to apply one or the other may be induced by the problem's frame, which means that the option for a strategy is a function of the problem's specificity. This is how humans react, in an implicit or explicit way, depending on the problem's evolution.

In some sense these two forms of human behaviour can be emulated by artificial entities, the software agents. Entities that have the ability to plan, to establish their actions ahead of time, to develop appropriated problem's solving strategies, to communicate, or to share resources. In a reactive system agents have the possibility to follow events as they occurs in the environment [6]. To be an effective task force, agents must also to cooperate. Sharing results or sharing tasks, are two common ways of doing it. In result sharing, agents work on interrelated sub-problems, interpreting and sharing knowledge or data. Agents do not act singly. The results obtained by a specific one may be used by the others. In task sharing, the problem is fragmented into sub-problems and distributed by a group of agents. Task distribution is done according to a prior deal in order to decide the allocation of tasks. Once task distribution is done, each agent work per se [9].

In this paper it is not intended to discuss the advantages/disadvantages of such models of behaviour or even to evaluate the different forms of cooperation among agents. Many of the ideas referred to above were adopted in the design and development of a distributed environment for the implementation of multi-agent systems: the **BeAble** system [1][4][5]. Integrated in this environment coexist different agent's classes. One of them is the expert one. Their architecture, dynamics, models of behaviour and cooperation will be presented. In the sequent a brief description of the distributed environment, its structure and global characteristics will be given.

2 The Distributed Environment

The **BeAble** system is a particular kind of an integrative environment for cooperative and distributed problem solving. It allows for the dynamic integration of several agents in an unique environment.

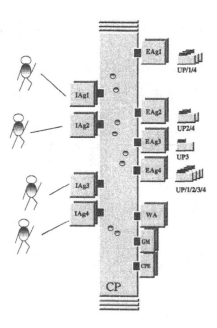

Fig. 1. The System's Environment

Agents can be full-scale expert systems, simple computational platforms to support interaction between users and the system or even computational entities emulating, for example, physical devices like sensors. The system's environment is made accessible to all entities. It is also dynamic in the sense that its status can change during the agents' activities. The system's dynamics is reinforced due to the possibility of activation/deactivation of agents at any time by its sponsors. Control is distributed by all system's components, according to their liabilities, as a way to avoid contention and deadlock situations.

The **BeAble** 's environment is illustrated in figure 1. One may observe several entities around a single common structure: the *Communication Platform* (CP). The CP is a vital system's component that caters for inter-agents communication and maintains system's global data. It is also used as a meta-knowledge structure, keeping information to be shared by the system's agents. In some aspects the CP resembles a typically blackboard data structure [8]. Indeed, high levels of knowledge organization, dynamic and flexible control, and the ability to endorse diverse problem solving strategies, are some of the relevant features of the blackboard paradigm that were integrated in the **BeAble** 's environment. In particular, providing the system's agents with opportunistic conducting, it was possible to improve effectively their dynamics and enhancing significantly their (asynchronous) behaviour.

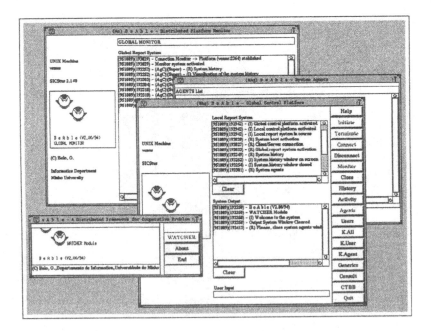

Fig. 2. The Watcher's Windows

The *Watcher* (Wa) module is another crucial system's component. It activates/deactivates the CP and initiates/terminates global system's operational tasks. Several tools for system administration are also available in the Wa's environment (figure 2), providing for systems monitoring, control and management of the agents' activities. The *CP Engine* (CPE) and the *Global Monitor* (GM) are Wa's support programs. The CPE maintains the CP active, supporting consequently the virtual structure of the distributed environment. The GM allows for the system's administrator to keep track of the agent's actions. It acts as a software demon looking continuously for new events. The remaining system's components in figure 1 are the *Expert Agents* (EAg) and the *Interface Agents* (IAg). The first ones are the system's experts. Skills, knowledge and resolution strategies of human experts are emulated through EAg, which present an expert system structure combined with specific protocol mechanisms to ensure communication with the underlying computational environment. The second ones are typically interface agents, used to assist users; i.e. providing for a medium that supports interaction between the users and the EAg's community.

The CP acts as a direct communication channel between agents. The system's agents communicate through the CP exchanging messages. Senders and receivers have local mechanisms that interpret the messages ensuring mutual understanding. The possibility for agents having their own language it is not put aside. In these cases, to ensure message interpretation and understanding agents must have adequate translation mechanisms. The message passing language is quite simple. It rules agent's developing, providing the structures and means used on each message type. Below are given three examples of messages, namely with respect to a query, a reasoning path request and a solution's interpretation:

$$bb(EAg,IAg,t\_eag\_question(EAg,IAg,Pr,NPr,Qu,Lv))$$
$$bb(IAg,EAg,t\_iag\_how(EAg,IAg,Pr,NPr))$$
$$bb(EAg,IAg,t\_eag\_method(EAg,IAg,Pr,NPr,Me))$$

where bb stands for a blackboard tuple; EAg and IAg, being respectively, the expert and interface agents; Pr is the code of the property under analysis being NPr its name; Qu is the text related with the question sent by the EAg; Lv is a list that contains the meaningful data for the query-answering; and Me is the text corresponding to the solution's interpretation.

The **BeAble** system was implemented in SICStus Prolog [7] under X Windows, and runs on a TCP/IP based network of Unix workstations.

3 The System's Experts

The EAg are the system's experts, whose skills and knowledge are bound to specific application domains. In general, their structure is quite similar to that of a conventional expert system shell, although complemented with mechanisms required for agent-to-agent communication. These are independent and autonomous entities, with their actions oriented by its internal state in conjunction with global CP information data-structures and by their reactiveness.

When active and connected to the system's CP, EAg act as asynchronous entities, running continuously, trying to detect on the distributed environment relevant knowledge. The EAg presents reactive behaviour, which allows for the agents to reason and act over their own state of affairs. Cooperative activities, involving result sharing or control passing, are also prerogatives of the EAg. This kind of social ability is critical when a problem requires a multi-disciplinary approach.

3.1 The Functional Architecture

An EAg is typically an expert system shell. It only assumes its role after reading a text file provided and maintained by the knowledge engineer: the knowledge base, which adopts a different role for a different knowledge base. The knowledge base is incorporated in the EAg structure only and when it is activated, and organizes itself into layers as follows:

- **agent characterization**, that asserts the EAg's independence and individuality;
- **acquaintance**, that keeps information related with others EAg in order to get help or advice, or even to pass (inference) control;
- **support information**, that includes data related with the EAg's domain, namely attributes, values, justifications and other raw data;
- **declarative rules**, that make up the EAg's inference engine.

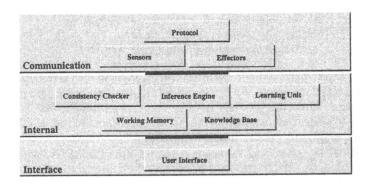

Fig. 3. The EAg's Architecture

The EAg's functional architecture as presented in figure 3, comprises:

- **A Communication layer**, with the contents:
  - Protocol mechanisms, that ensure communication between the EAg and the CP;

- **Sensors**, that operates on CP's data or knowledge relevant (that results from a cross checking between the user's *Agent Activation List* (AAL) and the support information contained in the EAg's knowledge base, being stored in the working memory) to the EAg's inference processes;
- **Effectors**, that place into the CP the results of the EAg's internal actions.

- An **Internal layer**, which is made up of:
  - **Inference Engine** (IE), that operates on the knowledge base and produces results, according to the principle of the "best candidate rule", based on the information provided by the Learning Unit;
  - **Working Memory** (WM), that stands for itself;
  - **Consistency Checker** (CC), that verifies and ensures consistency of the EAg's knowledge and data (every time a new property is added to or an old one has its value changed in the WM;
  - **Learning Unit**(LU), that improves rule selection based on the EAg's past experiences;
  - **Knowledge Base**(KB), that stores information about the application domain, knowledge, data and control information.

- An **Interface layer** - The bottom one contains a single software module, making possible the dialog with the EAg's administrator(s).

EAg integrate both reactive and deliberative behaviours. The former is achieved in the first layer by the sensors, which react to CP's changes induced by the EAg behaviour or data. Sensors' activity is supported directly by the WM. The knowledge collected onto the CP by the sensors and relevant to the EAg, is cross-checked with the one stored on the WM. Based on this comparison, the EAg detect new or updated data. The deliberative behaviour is achieved by meta-rules incorporated in the same layer for action selection and management and to support cooperation among EAg.

The functional structure just referred to above aims at a generic EAg in order to simplify software maintenance and generalize EAg generation. The selected inference mechanisms and knowledge representation schemes provide the EAg with adequate means to deal with diagnosis, classification or control applications, just to name a few. At least, modifications to EAg knowledge base can be made independently of any other EAg functional module. However, it is clear to us that the generality of the expert agent model may be engaged by the needs of some application domain.

3.2 Dynamics

The basic activity cycle of an EAg involves four main phases: *perception, analysis, action* and *communication of results*. In the perception one, an EAg try to detect on the system's CP data or knowledge that can be useful for its internal processes. This operation is done in two stages: 1) the EAg consults the cooperation structures (the users' AAL) that are involved with their users processes, trying to detect changes to the properties contained in the AAL; and 2) the EAg verifies if there are some messages sent directly to itself. To each user process active on the system's environment correspond always an AAL labeled with the user identification and time stamped. The analysis phase (figure 5 presents the procedure for the AAL analysis and consequent actions) is the one where the EAg decide what to do, based on the knowledge stored on its local WM and knowledge base. The CC is the first module to be activated, verifying consistency of the new or updated data confronting it with its previous values. If no consistency breakdown is detected then the IE is activated. It try to establish a new inference stage for the user's process referred to in the AAL. Upon termination of the IE, the LU verifies the current resolution strategy, analysis the status of the rules, calculate the new rules' weights and generate a new strategy (if it have the necessary knowledge and data to do it). The last phase is the communication of the EAg's results. These (if any) are posted by the effectors on the system's CP, addressed to the correspondent agents. The referred tasks can be followed through the EAg's local monitors (figure 4).

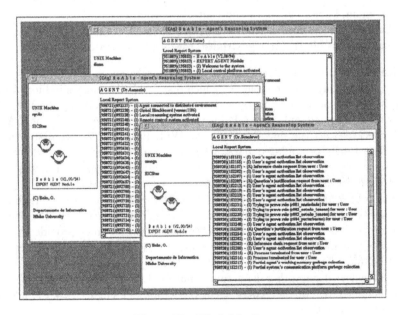

Fig. 4. The EAg's Monitors

3.3 Cooperation

In order to solve a problem EAg may use two different forms of joint operation. They may share intermediate or final results or use direct requests to pass inference control to others EAg. The AAL, located on the CP, are the knowledge structures that support the first form of action. The EAg use it to verify

```
procedure EAg-AAL-Analysis
    get-users-AAL-onto-CP(AAL-List);
    identify-users-on-local-active-processes(Users-List);
    select-active-users-on-AAL(AAL-List,Users-List,EAg-AAL-List)
    for each user-AAL(User-id,Properties-List) in EAg-AAL-List do
        identify-new-data-on-user-AAL(User-id,Properties-List,New-Data);
        select-from-new-data-relevant-items(User-id,New-Data,Relevant-Items);
        consistency-checking(User-id,Relevant-Items,Consistency)
        repeat
            if Consistency then
                integrate-new-relevant-items-on-WM(User-id,Relevant-Items);
                apply-inference-mechanisms(User-id,Results);
                integrate-results-on-WM(User-id,Results);
                update-user-AAL-on-CP(User-id,Results);
                monitoring-events(User-id);
                set Termination to true
            else
                conflict-resolution(User-id,Consistency)
                if not Consistency then
                    terminate-user-process(User-id);
                    broadcast-event(User-id);
                    monitoring-events(User-id)
                else
                    set Termination to false
                end-if
            end-if
        until Termination
    end-for
end-procedure
```

Fig. 5. The EAg's AAL-Analysis Procedure

property changes, or as an information "vehicle" for their own contributions to the user's process evolution. On a cooperative process, all EAg involved update continuously the correspondent user's AAL structures with new property's values or even with final results. Thus, the EAg have permanent access to the new problem states and consequently have the possibility to use the available information to evolute by itself on the user's process. When a specific EAg do not have the necessary ability to continue locally the inference process, it have the means to pass control of the inference process to another EAg. Such information is located, when available, in the acquaintance part of the EAg knowledge base and it is provided during knowledge base construction by the knowledge engineer.

3.4   Consistency Revision

Once the system allows agents to deal with dynamic knowledge (properties that change their values as time past during EAg interventions) a strong consistency revision policy was defined. Every time an EAg detects a property change (a new or updated property's value) on the user's AAL it activates the CC. The CC compares the pairs (property,value) contained on the AAL with the ones on the EAg's WM in order to detect the contradict ones. If contradictions are detected, the EAg stores the current inference status and applies its inference mechanisms (based on the new acquired data) to detect eventual consistency breakdowns. If this is the case the EAg reports the detected breakdowns, discharge the previous inference's steps, and broadcast the event to the system's EAg involved with the user's process. Upon message's reception the EAg suspend all the user's processes, and evaluates the impact of the consistency breakdown on their local bases, and plan to restore consistency (globally and locally). If consistency is restored the user's AAL is updated and the inference process restarted, and the event is broadcasted to all EAg. Otherwise the processes terminate, and the user is informed [2]. One may therefore claim that conciseness and consistency in the EAg's knowledge bases and working memories are preserved.

3.5   Learning Abilities

In an EAg knowledge base one may find two types of rules: the dynamic and the static ones, with the formers entitled to change their call orders in the inference process. One may ask: why to create two different types of rules ? Imagine, for instance, that a physician receives at a clinic, during a couple of days, a considerable number of patients, having detected flue in most of the cases. Will not be valid that the physician will analyze, in the next patient, the specific case of the flue with another kind of strategy and priority? The necessariness to obtain a problem's solution in a short period of time can eventually raise such kind of questions.

The EAg's LU was developed based on such frame of mind. It allows for the EAg to have the possibility to apply different resolution strategies according to the cases that were managed before. It can "learn" with past cases. Supported by such data it can reach a problem's solution in a shorter period of time. The LU works as a typical rewriting system. It takes the knowledge base rules as input and based on previous inference steps (triggered rules and the number of proved antecedents), calculates new weights for the rules (only for the dynamic ones), and rewrites their trigger's conditions. This operation establishes a new criterion of rule selection for the next inference processes. Implicitly, this process defines new rules' priorities on the IE selection process, helping the IE to short selection times at the rule conflict set. The dynamic application of different resolution strategies can optimize and reduce effectively EAg's rules analysis and consequently its resource allocation.

```
procedure Query-Answering
     Collect EAg's queries onto the CP
     Enqueue the queries according their times of arrival
     for each query(EAg,Message) do
          Dequeue(query(EAg,Message))
          if EAg's query already answered then
               Dismiss query
          else
                    present-query-to-user(EAg,Message)
                    wait-for-the-answer(User-Answer,Parameters)
                    validate-the-answer(User-Answer,Parameters)
                    case User-Answer of
                         value : Update AAL with (User-Answer,Parameters)
                                   on CP and on IAg's WM
                         why   : Send bb(IAg,EAG,t_iag_why(IAg,EAg,Message))
                                   and waits for EAg answer
                         how   : Send bb(IAg,EAg,t_iag_how(IAg,EAg,Message))
                                   and waits for EAg answer
                         status: Consults IAg Working Memory and shows
                                   the problem solving status
                         abort : Send bb(IAg,EAg,t_iag_abort(IAg,EAg))
                                   to all EAg involved with the IAg process
                                   Re-starts IAg Inference Process
                         quit  : Send bb(IAg,EAg,t_iag_end(IAg,EAg))
                                   to all EAg involved with the IAg process
                                   Terminates IAg Inference Process
                    end-case
          end-if
     end-for
end-procedure
```

Fig. 6. The IAg's Query-Answering Procedure

4   Using the System

System's experts are accessible for consulting using an IAg platform. The IAg act as a gateway allowing EAg access. Inside the IAg's environment users can apply for different services, which include information about EAg's application domains, skills or knowledge, the list of the current EAg available or how to request access to the system's EAg community. With respect to the EAg consulting services, users are presented with two choices. They may opt by a single or by a team of EAg. In the first option consultation is naturally restricted to the EAg's application domain, its expertise and knowledge. In the second option, the users have the possibility to be attended simultaneously by a group of EAg. After selecting the consultation mode, users have access to the IAg's inference environment, being able to interact with the EAg. The basic IAg life-cycle, related with EAg's query-answering, is depicted in figure 6.

5 Conclusions and Future Work

The present work endorses a distributed environment for multi-agent systems based applications, with an emphasis on the EAg, their dynamics and models of behaviour and cooperation. The system's dynamics allows for the possibility of dynamic integration of agents on the system's environment, at run time, by its local administrators. Moreover, the distribution of control by all the system's entities, according to their liabilities and abilities, provides a better way to improve system's flexibility and agent's autonomy, reduces contention and deadlock situations and facilitates software maintenance. The use of a fixed and dedicated structure for agent communication has revealed enormous advantages and it is a privileged mean to develop cooperation among agents. The analysis of the actual implementation suggests a redefinition of the model that supports the actual prototype system, towards a more efficient and robust one. The experiences realized and the testbeds developed (a medical diagnosis application and a simulation model of an air-conditioned distribution network [3]) have revealed the necessity of further research that points to: ensure robustness and flexibility on EAg reasoning processes; augment the expressiveness power of the EAg's knowledge representation; improve EAg learning abilities; and to design and develop a model for cooperation among different multi-agent communities with distinct environments.

References

1. O. Belo. Beable: A distributed framework for cooperative problem solving. Technical report, Departamento de Informática, Universidade do Minho, October 1994. (In Portuguese).
2. O. Belo and J. Neves. Consistency revision in a distributed multi-agent system environment. In *DIMAS'95 International Workshop on Decetralized Intelligent and Multi-Agent Systems*, Krakow, Poland, November 1995.
3. O. Belo and J. Neves. Multi-agent systems based distributed intelligent simulation - a case study. In *Eurosim Congress '95*, Vienna, Austria, September 1995.
4. O. Belo and J. Neves. A prolog implementation of a distributed computing environment for multi-agent systems based applications. In *Proceedings of The Conference of The Practical Application of Prolog (PAP'95)*, pages 31–41, Paris, France, April 1995.
5. O. Belo and J. Neves. A distributed problem solving environment for multi-agent systems. In *Conference of The Third World Congress on Expert Systems*, Seoul, Korea, February 1996.
6. B.Hayes-Roth. Opportunistic control of action in intelligent agents. *IEEE Transactions on Systems, Man, Cybernetics: Special Issue on Planning, Scheduling, and Control*, 23(6):1575–1587, 1992.

7. M. Carlsson and J. Widen. *SICStus Prolog User's Manual.* Swedish Institute of Computer Science, January 1993.

8. R. Engelmore and T. Morgan. *Blackboard Systems.* Addison-Wesley Publishing Company, Inc., 1988.

9. R. Smith and R. Davis. Frameworks for cooperation in distributed problem solving. In A. Bond and L. Gasser, editors, *Readings in Distributed Artificial Intelligence,* pages 61–70. Morgan Kaufmann Publishers, Inc, 1988.

A Scenario-Based Design Method and an Environment for the Development of Multiagent Systems

Bernard Moulin and Mario Brassard
Laval University, Computer Science Department
and Research Center on Geomatics
Pavillon Pouliot, Ste Foy, QC G1K 7P4, Canada
ph: (418) 656 5580, fax: (418) 656 2324; Email: moulin@ift.ulaval.ca

Abstract
We present the main modelling steps of the MASB method, an analysis and design method for the development of multi-agent systems viewed as systems composed of software agents playing various roles in predetermined scenarios. We show how agents can play several roles at once and how roles and scenarios can be used to partition agents' knowledge bases. We describe specific roles such as a scenario manager, an object server and a conversation interpreter which are implemented in SMAUL, an environment which is used to generate the main modules of scenario-based multi-agent systems. We present several modelling techniques that can be used in collaboration with users to describe scenarios, agents' roles and their main knowledge structures during the analysis phase. During the design phase these models are transformed by designers into formal specifications as inputs to SMAUL.

1. Introduction

Researchers and industry are actively developping *software agents* (SAs), autonomous software (ACM 1994, IEEE 1995) that will assist users in achieving various tasks, collaborate with them in order to solve problems, or even act on their behalf (Pan and Tennenbaum 1991, Maes 1994). This trend benefits from research works (Jennings and O'Hare 1995, Lesser 1995, Moulin and Chaib-draa 1995) done in *distributed artificial intelligence* (DAI) and *multiagent systems* (MASs). While SAs will become more widely available, developers will be willing to develop applications of increasing complexity based on the joint activities of groups of SAs. At that point they will face a methodological difficulty: how might they analyse an application domain in order to identify relevant SAs, to model their knowledge structures and behaviours, as well as their interactions with users? The specialized literature contains very few papers proposing methodological guidelines for developing MASs and systems composed of several SAs (Moulin and Cloutier 1994, Kendall et al. 1995. When it comes to MASs and SAs, we think that it is important to consider methodological issues from a user's point of view (Moulin et al. 1992) (Norman and Draper 1986), (Schuler and Namioka 1993). An analysis and design method for MASs should be based on concepts that are simple and familiar to users and yet powerful enough to model interaction patterns involving autonomous agents as well as the main knowledge structures that they need to manipulate. The agent models currently proposed by researchers are far too complex to be used by a user or a designer who is not specializing in DAI. See for instance (Cohen and Levesque 1990, Rao and Georgeff 1995) and various other models presented in (Jennings and O'Hare 1995) and in (Lesser 1995).

It is easy to explain users how SAs interact in a MAS using an analogy drawn from a theatrical world in which actors play roles in scenarios. Using this simple idea we elaborated the first version of the MASB method (MASB stands for "Multi Agent Scenario-Based"), an analysis and design method for the development of MASs and systems composed of several collaborating SAs in which agents are thought of as characters playing various roles in predefined scenarios (Moulin and Cloutier 1994). In this paper we present a new version of the MASB method which addresses the needs (requirements) of both users and designers of a scenario-based MAS. We also present the main characteristics of the SMAUL2 environment (SMAUL stands for "Systèmes Multi Agents Université Laval") which is used to model MASs on the basis of the MASB method specifications. The SMAUL environment is implemented in BNR Prolog on Apple MacIntosh workstations. Like most system development methods, and specifically object-oriented methods (Carmichael 1994), the MASB method is composed of three phases: analysis, design and implementation. The analysis phase involves both users and designers and aims at describing the application from a user's point of view. The design phase aims at transforming the application descriptions from the analysis phase into specifications that completely characterize the system architecture, scenarios and agents' models. The implementation phase transforms the design specifications into system programs. In this paper we concentrate primarily on the analysis and design phases. Fig. 1 provides an overview of the method.

Section 2 presents the main steps of the analysis phase: we discuss the use of scenarios from a user's point of view; we also present techniques that are used to provide functional descriptions of scenarios and to model relevant static and dynamic data structures. Section 3 presents the main knowledge structures used to model SAs (mental states, decision and action spaces, scenarios and roles) as well as the architectural properties of a MAS. Section 4 presents the main steps of the design phase, mainly how to model SA's belief structures, decision and action spaces.

2. The Analysis Phase

2.1 Modelling Scenarios: a User's Point of View

In AI the notion of scenario (or *script*) was used to specify behavioural patterns of planning systems (Schank et al. 1977). Recently, several authors working on human computer interactions (SIGCHI 1992) and object-oriented methods (Jacobson et al. 1992, Wirfs-Brock et al. 1990) promoted the idea of using scenarios as an informal way of describing an application in collaboration with users (Carroll 1995). In these approaches a scenario is viewed as a narrative, a graphical or a visual description of what people do when they perform particular tasks. Usually scenarios are thought of as intermediate descriptions from which system specifications can be derived.

In the MASB method scenarios are not only used to describe agents' interactions but also to structure the knowledge used to specify agents' behaviours. Users are involved in "storyboarding activities" during the analysis phase in order to describe SAs' interactions in the context of scenarios that characterize the application at hand. Then, during the design phase designers refine users' scenario descriptions and formally specify SA's behaviours and knowledge structures used to implement the MAS.

A user can imagine a MAS as a team made up of human and artificial agents. Artificial agents are linked together by a telecommunication network and mediate human interactions. Human and artificial agents interact together through communication interfaces supporting "conversations" in which the initiative can come from users or SAs. Agents not only interact with one another, but they also have access to data bases containing information that represent the world in which they evolve. This world contains artefacts (documents, objects, machines, etc.) created or manipulated by agents. Artefacts can also evolve independently according to the laws applying to the world (for example the gravitation law or market laws or social laws).

In the first step of the analysis phase designers help users to describe typical scenarios in a textual form, possibly augmented with some diagrams or sketches. This textual description may be more or less accurate. It is a starting point for storyboarding activities. In order to illustrate our presentation, we use the practical case of the development of a meeting scheduling system in which SAs negotiate together on behalf of their users in order to find a suitable time slot for a meeting allowed by their users's agendas. Since scheduling a meeting is a negotiation activity, users and designers described a scenario based on a negotiation protocol inspired by Cammarata et al. (1983). The negotiation takes place in several steps briefly described hereafter.

Analysis Phase
- A1 Scenario description (using natural language)
  . first identification of important notions supporting users'   scenarios:
  agents' description, agents' roles, objects, agents' interactions, object changes...
- A2 Role functional description (behaviour diagrams)
  . validation of agent interactions in relation to the scenario
- A3 Data conceptual modelling
  . local data modelling
  . object life cycles
- A4 Static and dynamic descriptions of the world
- A5 System-user interaction modeling
  . user/agent conversations (connected speech acts, interface specification, etc.)
  . validation of conversations in relation to the scenario.

Design Phase

- D1 MAS architecture and scenario characterization
  . identification of agents of different types (object servers, interpreters, SAs)
  . identification of scenarios and sub-scenarios
  . distribution of roles among SAs (refine roles into sub-roles)
  . distribution of functionalities among agents (first version of goals and plans)
- D2 Object modeling
  . object structure specification (attributes, relations, inheritance, procedural attachments)
  . object life cycle
  . object behaviour
  . validation of object/agent interactions in relation to scenarios
- D3 Agent modeling
  . characterization of the belief structures (including belief state change) for each agent;
  . modelling of the decision space for each agent
  . modelling of the action space for each agent
  . validatation those specifications against scenarios
  . factoring of the decision spaces
- D4 Conversation modeling
- D5 System design overall validation
  . when conversation modeling (A3) is completed, simulate complete scenarios
  including SAs behaviours and user/Sa interactions with the user

Fig.1: An overview of the analysis and design phases of the MASB method

---

Initiation phase. A user asks her SA (called Initiator-SA) to organize a meeting with other persons. Initiator-SA contacts the SAs of the potential participants and proposes to them that their users meet at a given period. Each potential-participant-SA contacts its user and asks her if she desires to participate in such a meeting. It then informs Initiator-SA about that decision. Initiator-SA gathers the answers of potential-participant-SAs, creates the participant list and sends it to the participant-SAs: this new role is assigned to the potential-participant-SAs who accepted to participate in the meeting as well to the initiator-SA for the following step of the negotiation.

Role determination. When it receives the participant list, each participant-SA establishes a "constraint factor" measuring the availability of its user during the period proposed for the meeting. The constraint factor is computed using the time slots' values ("free", "displaceable" or "fixed") recorded in the user's agenda for that period. Each participant-SA sends its constraint factor to every other participant-SA. After the exchange of constraint factors, each participant-SA can establish an ordering of the participant-SAs on an "availability scale"; hence, the new roles of participant-SAs: most-constrained participant (MCP), second-most-constrained (2MCP), etc.

Negotiation rounds. The negotiation is led by MCP using the following strategy: it will first propose to hold the meeting during time periods which correspond to "free" time slots in its agenda. The first "free time slot" which is suitable to every participant-SA is chosen. In order to limit communication exchanges, agents negotiate two by two. MCP proposes a meeting slot to 2MCP. If 2MCP cannot accept the proposed slot, it immediately notifies MCP who will have to make another proposal. If 2MCP agrees with the proposal, it transfers it to 3MCP. If 3MCP cannot accept the proposed slot, it immediately notifies 2MCP who transmits the refusal to MCP. This negotiation chain is applied from the most constrained participant to the least constrained participant. If an agreement on "free time slots" is not reached , MCP considers "displaceable" time slots in its agenda. Proposals are also transmitted along the negotiation chain. When transmitting its answer, each agent also indicates its constraint factor for each proposed period. MCP memorizes all the constraint factors in order to determine the best compromise for the group of agents.

Negotiation termination. If the negotiation is successful, MCP sends the meeting date and time to each participant-SA who updates its agenda and informs its user. If the negotiation is unsuccessful, MCP notifies to initiator-SA that no solution is available given the current constraints. Initiator-SA indicates to its user that the negotiation has been unsuccessful and asks if some constraints may be relaxed. If some constraints are relaxed by the user (such as the meeting period of potential participants), the negotiation is reactivated at the role-determination phase.

2.2 Scenario Functional Description

A textual description of scenario should emphasize the roles played by human and artificial agents, the typical information exchanges, events that occur in the course of the scenario and the actions performed by agents. In a complex MAS users and designers may identify several scenarios. If the scenarios are interrelated in some way, the description must emphasize those interrelations. An agent can play several roles in one or several scenarios. Certain situations may induce role changes as in our meeting scheduling scenario (Sect. 2.1) in which SAs can play several roles: "initiator", "potential-participants", "participants", "role evaluator", "MCP", etc.

In the MASB method we use *behaviour diagrams* (Moulin 1983, Moulin and Cloutier 1994) that enable a designer to precisely describe the scenario in terms of agent roles and activities (processes), local information or knowledge stores (accumulations) and their contributions (channels) to activities, and interactions (flows) with other agents (environments) playing specific roles. Behaviour diagrams are validated with respect to the scenario textual description (Sect. 2.1), which may be refined and updated as a result of this validation.

A behaviour diagram is associated with a role played by an agent: it is a tripartite graph whose nodes are called *processes, accumulations* and *environments* and linked by relations called *flows* and *channels*. Processes may be refined into more detailed behaviour diagrams. Graphically speaking, processes, accumulations and environments are represented by rectangles, rectangles with rounded corners and squares with a double outline respectively. Channels and flows are respectively represented by thick lines and arrows. Processes related to a specific behaviour are embedded in a rectangle which represents the corresponding behavioural plan. Fig. 2 presents two behaviour diagrams corresponding to two roles that an agent (called AMSA for Agenda Management Software Agent) can play in the meeting scheduling scenario.

220

Diagram 1 describes the behaviour plan *Initiate-meeting-scenario* associated with the *Initiator* role. A *user* (E0) sends her agent a *meeting-request* that is received by process P1.1, which registers the *Meeting-information* (A1): subject, duration, earliest and latest dates, and persons to be contacted. Process P1.2 sends a *meeting-proposal* to the *Potential-participants* (E1) according to the list of invited persons. *Potential participants* (E1) send back their *answers* that are received by process P1.3 which creates the *Participant-list* (A3). When all the agents' answers have been received, the *Initiator* agent sends the *participant-list* to the selected *Potential-participant* agents. Diagram 2 describes the behaviour plan *Reply-to-meeting-proposal* associated with the *Potential-participant* role. Other SA's roles are described in (Moulin Cloutier 1994).

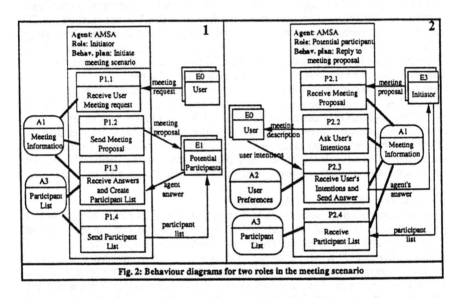

Fig. 2: Behaviour diagrams for two roles in the meeting scenario

Behaviour diagrams are easily understood by users because they present a functional description of the system's behaviour. They can be easily checked against the textual description of the scenario: users and designers can play the various agent roles in a simulated scenario on the basis of the behaviour diagrams and make sure that all desired agent functionalities are included in them. Behaviour diagrams are more generic than time-line diagrams (Rumbaugh et al. 1991, Jacobson et al. 1992). They can also be used to formalize the information obtained from CRC cards (Wirfs-Brock 1990).

2.3 Conceptual Data Modelling
In the context of a given scenario, a designer creates a set of behaviour diagrams for each SA. The set of accumulations contained in these behaviour diagrams corresponds to the data and knowledge used by the SA. During the conceptual data modelling phase (step A3 in Fig. 1), a designer analyses the contents of these accumulations in terms of attributes and creates a conceptual data structure featuring the main elements that will likely be included in the SA's local data base. This conceptual data structure may be represented respectively by "entities and relationships", by "classes, sub-classes and relationships" or by "frames", depending on whether the agent's local data base is implemented with a relational data base, an object-oriented data base or a frame base. The details are discussed in (Moulin Cloutier 1994).

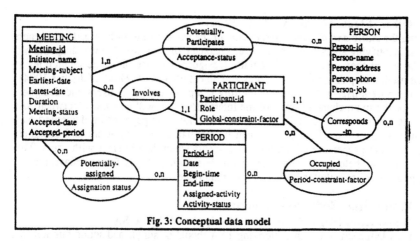

Fig. 3: Conceptual data model

The conceptual data structure specifies the static properties of an agent's fact base. At any given time the state of an agent is characterized by the content of its fact base. The agent's evolution is reflected by the value changes of its attributes. Not all state transitions are not allowed, however. In the MASB method, using object life-cycles (Moulin 1983), a designer conceptually specifies the allowed state transitions for the object.

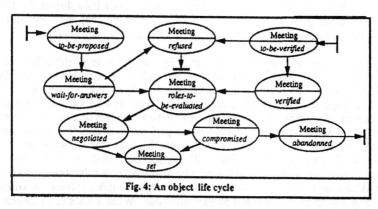

Fig. 4: An object life cycle

An *object life-cycle* diagram is a connected graph in which the nodes represent the object states, and the edges represent the transitions allowed between those states. The object life-cycle indicates which changes are allowed in the instances of a given object. Fig. 4 shows the life-cycle associated with the entity *MEETING*. The names of states are indicated in italic in the ovals. In the initiator-agent data base the meeting is created in the "to-be-proposed" state and becomes "wait-for-answers" until all the potential participants' answers have been received. The symbol I-> indicates the creation of an instance in the corresponding state. If all the potential participants refuse to participate, the meeting instance becomes "refused". In a potential-participant local base the meeting instance is created in the state "to-be-verified", and becomes "verified" (resp. "refused") if the user accepts (resp. refuses) to participate in the meeting. Then the meeting instance takes the states "roles-to-be-evaluated", "negotiated", "compromised", "set" or "abandoned" depending on the negotiation stage. The symbol ->I indicates that the instance can be deleted from the corresponding state.

2.4 Other Steps of the Analysis Phase

During the *static and dynamic description of the world* (Step A4 in Fig. 1), designers aim at identifying and structuring the data characterizing the world in which agents act: artefacts that they create or manipulate. Designers can use the same techniques as those used in step A3: object structures and object life cycles (Moulin,Cloutier 1994). During *system/user interaction modelling* (Step A5 in Fig. 1) users and designers describe the main characteristics of the interactions that will take place between SAs and users in the form of sketches of interfaces drawn on paper or implemented on a form editing system. The scenario can be simulated again by users and designers including the proper use of interfaces by agents when they interact with users. This simulation can help users to have a better understanding of SAs' future behaviours and designers make sure that they fulfill users' requirements.

3. Modelling Software Agents

3.1 Main Knowledge Structures Manipulated by a SA

The knowledge structures characterizing a SA should be simple and sound familiar to a designer. Fig. 5 presents the graphical formalism that we use to specify the knowledge structures of a SA. At any time a SA is characterized by its *mental states* such as beliefs, goals and expectations. A SA acts according to its mental states, to other SA's actions and to the changes occurring in the world in which it evolves. Its actions are specified by different kinds of *transition rules* that have *preconditions* (conditions that should hold in order to trigger the rule), *postconditions* (effects of triggering the rule) and *coconditions* ("concurrent conditions" that should hold during the time period when the action is performed). Preconditions, postconditions and coconditions may be composed of any combination of mental states and message structures.

A SA has *beliefs* (round-cornered rectangle in Fig 5.1) about the world: information describing agents and objects as well as situations. Beliefs are recorded in a SA's *fact base*. SAs communicate together by exchanging *messages*. Fig. 5.2 shows how to specify a message (round-rectangle over an arrow) emitted by a SA's action (rectangle) and directed toward another SA (double square). Certain messages correspond to speech acts (Searle and Vanderveken 1985) that SAs can perform (such as orders, requests and promises). Messages are also used to model events that happen in the world in which SAs evolve. *Temporal marks* (Fig. 5.7) are specific points in time (such as deadlines or particular dates) or time intervals (such as a delay or a duration allowed for an action) that may be significant for a SA. A temporal mark contains a predicate representing a specific situation and an expiration date. When a SA's clock indicates that the expiration date is reached, the temporal mark expires and triggers the actions for which it is a precondition. An *expectation* (Fig. 5.8) indicates that an agent expects that in a foreseeable future something will happen or an agent will act in a certain way. An expectation is also associated with an expiration date.

A SA considers *goals* (oval box in Fig. 5.11) that might permit it to reach desired states. A goal can be decomposed into sub-goals which provide detailed ways of achieving that goal. When a SA chooses to achieve a goal, an instance of that goal is created in the *active* state in its goal agenda. If that goal has sub-goals, instances of these subgoals can be activated. Depending on the circumstances, a goal may be achieved (it is said to be a *success*) or not (it is said to be a *failure*).

A SA knows about the goals it can work towards, thanks to a *decision space* (Lizotte, Moulin 1990) which is a structure composed of goal hierarchies augmented with relations between goals and beliefs. The activation of a goal might depend not only on a SA's intention of reaching it, but also on other conditions involving for example the existence of particular beliefs in the SA's fact base or the arrival of a particular message. Fig. 5.11 symbolically presents a portion of a decision space.

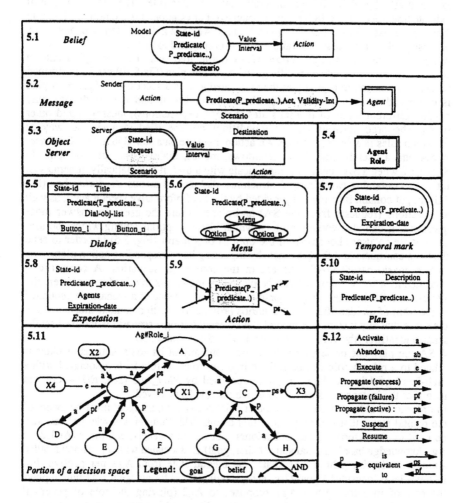

Fig. 5: Formalism for representing an agent's knowledge structures

The decision space gathers different kinds of transition rules used to model a SA's decision making process: an *activation rule* indicates how a goal can activate its subgoals; an *execution rule* features the condition that should be true in order to execute the goal; a *propagation rule* indicates how the success or failure of a goal is propagated to other goals; other transition rules specify how a goal can be abandoned, suspended and resumed. Graphically a transition rule is represented by a set of arrows linking the goals and beliefs and possibly the messages involved in the rule. Fig. 5.12 presents the main conventions used to represent these rules using different kinds of annotations on arrows. For example in Fig. 5.11, goal A activates goal B in conjunction with the mental state X2 (*arrows a* related by a segment symbolizing an

AND operator), while goal C is activated without any condition. When it is active, goal B activates goals D, E and F. In order to be executed goal B requires that the execution condition on belief X4 holds (*arrow e*), but then goal D, E and F will be executed as subgoals of B without additional execution rules. The *arrows ps* and *pf* indicate how success and failure are propagated. For example it is the failure of goal D that is propagated to goal B. For goals E and F we have a disjunction: whenever one of these goals is a success or a failure, this is propagated to goal B.

A SA knows about various *action plans* (structured sets of actions) that can be activated in order to achieve specific goals. An action plan (Fig 5.10) can be performed successfully or not, implying that the associated goal is a success or a failure. A plan is composed of *actions* (Fig 5.9) whose preconditions and postconditions may contain different kinds of mental states (beliefs, expectations), messages and other items. The preconditions and postconditions are specified in a similar way as transition rules in a decision space. An action may use data stored in a SA's fact base.

A SA and a user interact using windows, menus, dialog boxes and buttons. In SMAUL2 when a SA needs to communicate with a user, it automatically generates an interface containing two kinds of structures: dialogs and menus. A *dialog* box (Fig 5.5) might contain information provided by the SA (informative speech acts), as well as questions (requests from the SA) and places where the user can answer those questions. It can also contain buttons which identify options among which the user can choose. A user chooses from the options of a **menu** (Fig. 5.6) in order to send the SA a command and the SA reacts by triggering its corresponding action plan.

In a scenario-based MAS SAs act in the context of scenarios. A *scenario* is specified as a set of roles that can be played by the agents involved in that scenario. Whenever a SA initiates a new course of action, an instance of a scenario is generated in the form of an identifier *scn*. Scenarios provide SAs with a context mechanism in the sense that any instance of mental state created during a scenario is indexed by the senario identifier (the variable *scenario* in Fig. 5.1, 5.2 and 5.3).

A SA can play one or several *roles* at once and any of its goals, actions or plans is related to a role. A role can be refined into sub-roles. A role Ri is associated with a high-level goal in the decision space specified by *Play_role (scn, Ri)* where scn identifies the instance of scenario in which the role Ri is played.

In order to be able to reason about other SAs' behaviours, an agent must be aware of the mental states that it shares with other agents. This "mutual knowledge" can also be used to characterize group behaviours. Let Ag be a generic SA in SMAUL2. In Ag's knowledge base a mental state (belief, goal) is associated with a parameter called a *model* indicating which SAs share that mental state from Ag's perspective. For instance let us consider a goal *Organize_Meeting(Mng)* for an agent Ag1 which plays the role of initiator. When Ag1 proposes to Ag2 (playing the role of potential participant) that it participates in the meeting Mng, it is proposing that Ag2 adopts the goal *Organize_Meeting(Mng)*. If Ag2 accepts, Ag1 can deduce that it shares that goal with Ag2. That information is expressed by associating to this goal the *model parameter* "Ag1 •+ Ag2". The operator "•+" indicates that Ag1 believes that Ag2 has the goal *Organize_Meeting(Mng)* and that Ag2 believes that Ag1 has the same goal. As we have seen, an agent's knowledge base is organized according to the roles it plays. We use the operator # to specify the role that an agent plays. For example Ag1#initiator indicates that Ag1 plays the role of initiator. Since a SA is aware of its roles and of other SAs' roles, the model parameter applies on roles. Hence, the proper way of indicating that Ag1 playing the role of initiator and Ag2 playing the role of potential participant (pot.participant) share the goal *Organize_Meeting(Mng)* is expressed by the model parameter: Ag1#initiator •+ Ag2#pot.participant.

The worlds in which agents evolve and act are symbolized by *object servers* which contain the world knowledge shared by SAs. An object server contains objects that represent artefacts (documents, things, machines, etc.) created or manipulated by agents. *Objects* are described as sets of attributes. Objects can evolve on their own in the world, their behaviours being modelled by simple action plans that the object server activates when required. In order to get information about the world, SAs can send requests (double round-cornered rectangle in Fig 5.3) to the object servers.

3.2 A Scenario-Based Architecture for MAS

From a designer's point of view a MAS gathers agents located on one or several computers (called the nodes of the MAS) linked by a communication network. A MAS can be composed of several kinds of agents: object servers, scenario managers, conversation interpreters and software agents.

An *object server* is a specialized agent that behaves as a deductive database or a "knowledge service" (Pan et al. 1991). It manages a data repository, called an *object base*, composed of objects representing artefacts. Objects are simple reactive agents whose behaviour is triggered by changes in their states as a consequence of SAs actions or of other object state changes.

A *scenario manager* helps a designer to specify scenarios and directs SAs according to these scenarios at run-time. At each node of a MAS there is a scenario manager which coordinates the various scenarios taking place on the node. Technically, a scenario manager is implemented as a special kind of object server. A scenario manager monitors the activation or deactivation, suspension or resumption of scenarios, of SAs and of the roles they can play.

A *conversation interpreter* and a user can have complete conversations composed of the chaining of several interfaces through which user's speech acts are performed by clicking actions on different kinds of buttons or menus and SA's speech acts are performed by displaying different kinds of interface elements. Conversation interpreters are intermediaries between users and SAs located on the nodes of the SMA.

In SMAUL2 object servers, scenario managers and conversation interpreters are roles that can be assigned to agents in a multiagent application.

The functional characteristics of an application must be specified by a designer in the form of knowledge structures characterizing SAs that play specific roles in the scenarios of the application. When a scenario is activated, the scenario managers activate the corresponding agents' roles on their nodes. An agent can only act in the context of the roles that it can play in a given scenario. Hence, decision spaces are partitioned according to SAs' possible roles.

Fig. 6 presents a generic partition of knowledge structures in a SA. A SA can play several roles at once. The fundamental role of an agent is *to_be_a_SA* and all other roles are sub-roles of that role. The role *to_be_a_SA* is always active when a SA is "alive" on a node. A role Ri is specified by the goal *Play_role (scn, Ri)* at the top of a decision space. Such a goal is persistent (Schank and Abelson 1977) in the sense that a SA cannot reach it with success or failure as a result of its activities: the role (and the corresponding goal) can only be created or deleted when a scenario manager creates or deletes a scenario in which this role is active. The decision space can contain any kind of goal required to model the decision making process of a SA playing that role. Some of these goals can activate plans in the action space. The transition rules contained in decision spaces access the instances of mental states contained in the agent's fact base. The reasoning space gathers the deduction rules that a SA possesses about the causal evolution of the worlds in which it evolves.

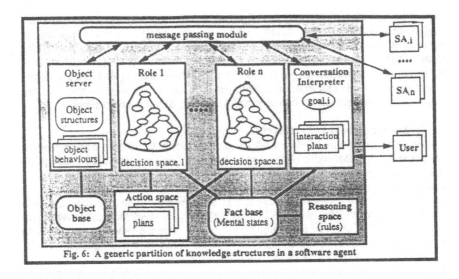

Fig. 6: A generic partition of knowledge structures in a software agent

A designer has to decide which SAs play the roles of object server and of conversation interpreter. Those roles are provided by the SMAUL2 environment. The designer has to specify to an object server the object structures and potentially the action plans associated with the objects. For a particular role of conversation interpreter the designer specifies the corresponding decision space (often limited to few high-level persistent goals) and the associated interaction plans mainly composed of dialog and menu knowledge structures. Several kinds of interactions can take place between roles. The decision space of a role Ri can contain a goal that instantiates a goal of the type *Play_role (scn, Rj)*,thus activating another role. Roles can also interact by exchanging messages. The message passing module (Fig. 6) manages message exchanges within the system: between roles of a SA or between SAs. Messages are specified as postconditions of actions. They are sent when the action is executed. When interacting with an object server, a SA sends a request to the object server which replies with the appropriate action on its object base.

4. The Design Phase

During the design phase (Fig. 1) the conceptual descriptions of a system obtained during the analysis phase are transformed into formal specifications. In this paper we present the different steps of the design phase in relation to the use of SMAUL2.

4.1 MAS Architecture and Scenario Characterization

Different questions must be addressed: which scenarios should be implemented in the MAS? Which roles should be involved in those scenarios? Which SAs should support which roles, including object server and conversation interpreter roles? Which functionalities should be assigned to roles played by SAs? During the *MAS architecture and scenario characterization* step (D1 in Fig. 1), designers answer those questions and specify the architectural characteristics of a scenario-based MAS. Users' scenarios provide a good starting point for this design step: in some cases they must be adapted by merging certain roles or splitting others, or by introducing certain agents. However, it is easy for designers to explain to users the rationale for such changes since users are already used to reasoning about scenarios and roles.

In our case study the scenario proposed during the analysis phase is kept unchanged for the design phase. Each user is assigned a SA called her *assistant* which will represent her during the meeting scheduling negotiations. For this scenario all assistant-SAs are identical and can play the same set of roles. Each assistant-SA plays the role of a conversation interpreter when interacting with its user. Each assistant-SA plays the role of object server when it comes to the manipulation of persistent data such as the agenda of its user. Each assistant-SA can play the different roles and that have been defined during the analysis phase (initiator, participant, MCP, 2MCP, etc.) according to its involvement in a given instance of the meeting scheduling scenario (Sect. 2.1). The behaviours described by the users using behaviour diagrams (Sect. 2.2) will be assigned to the appropriate SA's roles and specified properly during the agent modelling step (D3 in Fig. 1).

Object modelling step (D2 in Fig. 1) starts from the models obtained during the *Static and dynamic descriptions of the world* step of the analysis phase (A4 in Fig.1). During that step designers specify structures (class hierarchies, attributes, procedural attachements, etc.) and behaviours (plans) of objects contained in the object servers of the MAS under construction. When these specifications are complete, designers validate object/agent interactions in relation to scenarios. The next step of the design phase is *Agent modelling* (D3 in Fig. 1): designers specify the knowledge structures characterizing SAs: beliefs, decision space, action space and reasoning space.

4.2 Characterization of belief structures

The main elements of an *agent's belief structure* can be derived from the data conceptual structure obtained during the *data conceptual modelling* step of the analysis phase (A3 in Fig. 1). Designers must decide how to specify these elements as beliefs structures (Fig. 5.1) in short term memory or as objects (Fig. 5.3) manipulated by an object server. Usually an object server is used when beliefs must be recorded in long term memory. In our case designers chose to keep the information dealing with an SA's user agenda in the object base managed by the SA playing the role of object server. The information related to the current negotiated meeting is manipulated as belief structures in the SA's fact base. When the meeting is set, the user's agenda information is updated in the SA's object base. The main belief structures used by an assistant-SA is Meeting (Mng, meeting.state, participant_list) where Mng is a list gathering the information describing the meeting (entity MEETING in Fig.2)

Designers decided to simplify the life cycle of the MEETING object, keeping only five states. An instance of meeting is created in the state *initiated*. It becomes *negotiated* after the completion of the role evaluation step. After a successful negotiation the meeting instance becomes *set*. If the negotiation fails (no meeting period has been found), the meeting instance becomes *failed*. The instance of a meeting can also take the state *abandoned* if the user decides to withdraw her request for the meeting. These states will be used in the SA's decision space.

4.3 Specification of the decision space

Designers must specify an SA's decision space for each role it can play. Fig. 7 presents the main portion of an assistant-SA's decision space. For the purposes of the following discussion, each relevant element contained in the diagram has been assigned a number.

The upper goal is Play_role (Scn, Ag#agenda_man, scn_agent_lst) (1) and the model of this goal is Ag#to_be_a_SA (the basic role that any agent plays). This implies that any SA in our system adopts the goal of playing the role of agenda manager in the scenario Scn with a number of other agents (scn_agent_lst). This is a permanent goal.

This goal activates the goal Manage_user_agenda (2) when playing the role of agenda_conv_interpreter (as indicated in the associated model parameter Mdl0). This goal is another permanent goal. In turn it activates the plan pl_manage_agenda (3) which controls the interface through which the assistant-SA and its user interact (due to space limitation this plan is not presented in this paper): the user can consult her agenda and ask her assistant-SA to organize a meeting with other persons.

When receiving such a request, the SA Ag playing the role of agenda_conv_interpreter (4) sends a message (5) to itself playing the role of Agenda_manager. This message Propose_meeting (Mng, L_involved) includes the information about the meeting Mng and the list of persons proposed by the user. This message contributes to the activation of the goal Organize_Meeting (Mng, L_involved) (6) in the sub-scenario Scn°Sscn. When an initiator-SA starts a new meeting scheduling activity, it intantiates a sub-scenario Scn°Sscn which is used to mark in the mental models of the involved SAs all the interactions and mental states related to that activity: hence several meeting scheduling activities can be managed at once. Note that Organize_Meeting(...) is a group goal whose model parameter is Mdl1= Ag#initiator •+ L_involved#pot.participant, meaning that the initiator-SA Ag believes that it shares this goal with the SAs of the list L_involved when they play the role of potential participant. Note also that this goal can be activated when the agent Ag playing the role of potential participant receives a message Propose_meeting(Mng) (7) from another agent Ai playing the role of initiator (21). The goal Organize_Meeting (Mng, L_involved) (6) activates successively three goals Initiate_Meeting (Mng) (8), Eval_Roles(Mng, ...) (9) and Set_Meeting(Mng) (10): Organize_Meeting (...) is a success if Initiate_Meeting(...), Eval_Roles(...) and Set_Meeting(...) are successes and it is a failure if one of its sub-goals fails. Note that the success or failure of Organize_Meeting (...) is not propagated to its parent-goal Play_Role (...) (1). Instead, if Organize_Meeting (...) is a success, a message Meeting_set (Mng, L_participant) (12) is sent to the conversation interpreter (4). In the case of a failure the message Meeting_not_set (Mng, justification) (13) is sent to the conversation interpreter (4) who informs its user. The plan pl_manage_agenda (...) (3) is activated and the user can choose to change the meeting parameters or to abandon it. In that case the belief Meeting (Mng, ...) (11) takes the value *abandoned* and the goal Organize_Meeting (6) is abandoned.

The goal Initiate_Meeting (Mng) (8) activates the plan pl_initiate-meeting (Mng) (14) while the SA plays the role of initiator or of potential participant (see model parameter Mdl1). The success or failure of this plan is propagated to the goal (double arrow a-p). We will see in Sect. 4.4 that the outcome of that plan is to put the belief Meeting (Mng, meeting.state, participant_lst) into the state *initiated* and to create the list of participants accepting to join the meeting. If this list is empty (15), this is a failure (arrow pf) which is propagated to the plan header (14). If the list is not empty (16), this is a success (arrow ps) which is also propagated to the plan header (14). This belief Meeting(...) (19) in the state *initiated* triggers an execution rule for the goal Eval_Roles (Mng, [Ag#R_ag, R_roles..]) (9) which initiates the role determination phase of the scenario (Sect. 2.1). This group goal is adopted by all participant-SAs playing the role of evaluator (see model parameter Mdl2). This goal (9) activates within each participant-SA the plan pl_eval_role (Mng, [Ag#R_ag, R_roles..]) (17) which is used for the determination and exchanges of constraint factors and results in the identification of the roles of Most-constrained-participant (MCP), the second most-contrained (2MCP) etc. (Sect. 2.1).

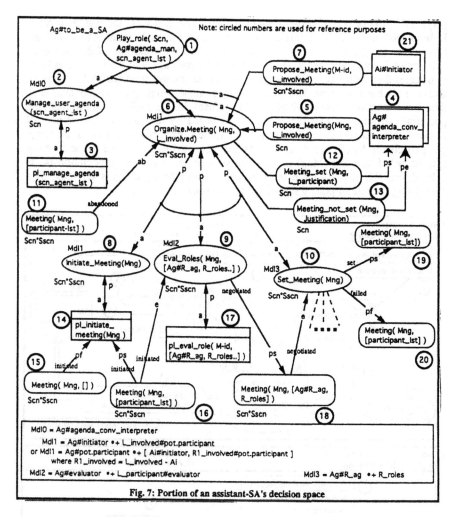

Fig. 7: Portion of an assistant-SA's decision space

4.4 Specification of the action space

Designers have to create the action space in which they specify the plans used by SAs. This step is usually done in conjunction with the creation of the decision space on the basis of the scenario descriptions obtained during the analysis phase. As an illustration Fig. 8 presents the plan initiate_meeting from two perspectives: that of a SA playing the role of initiator and of potential participant. This plan is activated by the goal Initiate_Meeting (Mng) (number (8) in Fig. 7). Circled numbers in Fig. 8 are used to refer to the relevant elements in the following discussion.

When the plan pl_initiate_meeting(Mng) is activated by a SA playing the role of initiator (Fig. 8.1), the first action to be executed is Ask_Meeting(Mng) (box (1) in Fig. 8.1) which sends a message Wants_Mng(Ag, L_involved) (2) to each SA playing the role of potential participant (3). In the action box (1) the line "F.a. Ak ∈ R_involved" indicates that this action is repeated for each agent Ak of the list R_involved (the set of involved SAs minus Ag). As a postcondition of that action an expectation Expect_ans_Mng(...) (4) is created for each agent of the list R_involved: it indicates that the initiator-SA expects to receive an answer from each SA to which a message (2) has been sent. The action Receive-Answ_Mng(...) (7) is activated each time that a message accepts_Mng(...) (5) or refuses_Mng(...) (6) is received from the potential participants (3): the answer is recorded by the initiator-SA. When all answers are received according to the agent's expectations (4), a message get_participant_lst (8) is sent to the SAs that accepted to participate in the meeting (3), the belief Meeting (Mng) is set to the state *initiated* and the expectation (4) is deleted. Fig. 8.2 displays the plan pl_initiate_meeting(Mng) which is activated by a SA playing the role of potential participant. It is not discussed here.

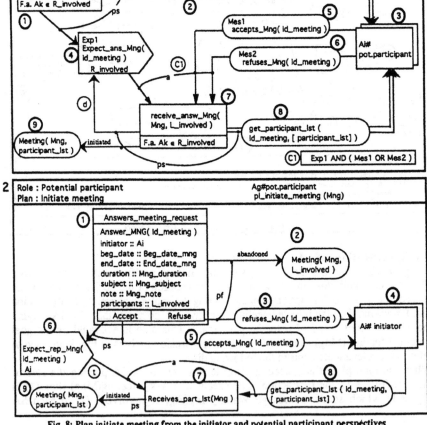

Fig. 8: Plan initiate meeting from the initiator and potential participant perspectives
Note: circled numbers are used for reference purposes

4.5 Conversation modelling and system design overall validation

During *conversation modelling* (step D4 in Fig. 1) designers specify the decision and action spaces of the SA playing the role of conversation interpreter. The corresponding plans mainly contain actions generating and controlling interface components such as dialog boxes (Fig. 5.5) and menus (Fig. 5.6). SMAUL2 automatically generates the corresponding interfaces. Using the specifications of the decision and action spaces SMAUL2 generates the programs that support the various SAs of the application. Designers may have to do some minor adjustments to the code. The *system design overall validation* step (D5 in Fig. 1) consists of simulating the application by triggering the behaviours of the various SAs playing the different roles included in the scenarios identified during the analysis phase.

5. Conclusion

In this paper we presented the main modelling steps of the MASB method, an analysis and design method for the development of multi-agent systems viewed as systems composed of software agents playing various roles in predetermined scenarios. Here are some remarks about our experience with the MASB method. As for any system development method, it is crucial to involve users and to provide them with modelling techniques that are easy to understand and powerful enough. Modelling scenarios and actors' roles using behaviour diagrams is such a technique. It is more general than use cases (Jacobson et al. 1992, Kendall et al. 1996) because it provides a functional view of the system integrating the notions of role, activity and memory. It is also important to provide designers with clear guidelines and models that are as simple as possible and yet expressive enough: the MASB method fits these requirements. Designers easily used the models of the analysis phase which sound familiar to them given their knowledge of information system or object-oriented concepts and modelling techniques. The models of the design phase are more complex, but the proposed concepts are easy to understand (belief, goal, expectation, temporal mark, action and plan). Designers had to be more careful with the different kinds of transition rules and with mutual knowledge. Another original and appreciated feature is the use of knowledge structures representing interactions with users in terms of interface components and their integration in SAs' action schemes and plans: this enables designers to specify interfaces in the same way as any other activity in the MAS. The integration of the notions of scenarios, roles, goals, beliefs and plans in the decision and action spaces provides a powerful model for specifying a SA's decision process in the context of the roles it can play in specific scenarios.

In the SMAUL2 environment designers appreciated the availability of specialized SAs: scenario managers, conversation interpreters and object servers. In future research we will use the MASB method to develop several kinds of multiagent systems. We will create new agent characteristics in order to support various kinds of reasoning mechanisms: hypothetical reasoning, modal reasoning, temporal reasoning. We will also explore the possibilities of using scenarios and roles to generate agents capable of reasoning about others and about their interpersonal relationships.

Acknowledgements

This research is supported by the Natural Sciences and Engineering Research Council of Canada (grant OGP 05518) and FCAR.

Bibliography

ACM (1994), *Communications of the ACM*, Special issue on Intelligent Agents, v37-17.

Cammarata S., McArthur D., and Steeb R. (1983), "Strategies of Cooperation in Distributed Problem Solving", *Proc. 8th Joint Conf. on AI*, Karlsuhe, Germany, 767-770, 1983.

Carmichael A. (edt) (1994), *Object Development Methods*, SIGS Books, New York.

Carroll J. M. (1995), *Scenario-Based Design*, Wiley.

Cohen P. R. and Levesque H. J. (1990), Persistence, intention and commitment, in Cohen P. R., Morgan J., Pollack M. E. (eds), *Intentions in Communication*, MIT Press, 33-69.

IEEE (1995), *IEEE Expert*, Special issue on Intelligent Internet Services, vol 10 n 4.

Jacobson I., Christersson M., Jonsson P. and Overgaard G. (1992), *Object-Oriented Software Engineering - A Use-Case Driven Approach*, Addison Wesley.

Jennings N., and O'Hare G. edts (1995), *Foundations of Distributed AI*, Wiley.

Kendall E. A., Malkoun M.T., and Jiang C.H. (1996), A methodology for developing agent-based systems, Proceedings of the First Australian Workshop on DAI, D. Luckose and C. Zhang edts., Lecture Notes on Artificial Intelligence, Springer Verlag.

Lesser V. edt.(1995), ICMAS-95, Proceedings of the First International Conference on Multi-Agent Systems, MIT Press.

Lizotte M., and Moulin B. (1990), A temporal planner for modelling autonomous agents, in Demazeau Y., Müller J-P edts, *Decentralized Artificial Intelligence*, Elsevier, 121-136.

Maes P. (1994), Agents that reduce work and information overload, in (ACM94), 31-40.

Moulin B. (1983), The use of EPAS/IPSO approach for integrating Entity Relationship concepts and Software Engineering techniques, in Davis C.G., Jajodia S., Ng P.A., Yeh R. editors, *Entity-Relationship Approach to Software Engineering*, North Holland.

Moulin B., Boulet M. M., and Meyer M.A. (1992), Toward user-centered approaches for the design of knowledge-based systems, in *Expert Systems for Information Management*, vol 5 n2, 95-123.

Moulin B., and Chaib-draa B. (1995), Distributed Artificial Intelligence: an overview, to appear in (Jennings et al. 1995).

Moulin B., and Cloutier L. (1994), Collaborative work based on multiagent architectures: a methodological perspective, in Aminzadeh F., Jamshidi M. edts, *Soft Computing: Fuzzy Logic, Neural Networks and Distributed Artificial Intelligence*, 261-296, Prentice Hall.

Norman D. A., and Draper S. W. (Edts.) (1986), *User-Centered System Design - New Perspectives on Human Computer Interaction*, Lawrence Erlbaum.

Pan J. Y. C. and Tenenbaum J. M. (1991), An intelligent agent framework for enterprise integration, in *IEEE Trans.on Systems, Man and Cybernetics*, vol 21, n 6, 1391-1408.

Rao A.S., and Georgeff M.P. (1995), BDI agents: from theory to practice, in (Lesser 1995), 312-319.

Rumbaugh J. , Blaha M., Premerlani W., Eddy F., and Lorensen W (1991), *Object-Oriented Modeling and Design*, Prentice-Hall.

Schank R., and Abelson R. (1977), *Scripts, Plans and Goals*, Lawrence Erlbaum.

Schuler D., and Namioka A. (Edts.) (1993), *Participatory Design: Principles and Practices*, Lawrence Erlbaum.

Searle J. R., and Vanderveken D. (1985) *Foundations of Illocutionary Logic*, Cambridge University Press.

SIGCHI Bulletin (1992), Special issue on scenarios, 24 (4).

Wirfs-Brock R., Wilkerson B., and Wiener L. (1990), *Designing Object-Oriented Software*, Prentice Hall.

Springer-Verlag
and the Environment

We at Springer-Verlag firmly believe that an international science publisher has a special obligation to the environment, and our corporate policies consistently reflect this conviction.

We also expect our business partners – paper mills, printers, packaging manufacturers, etc. – to commit themselves to using environmentally friendly materials and production processes.

The paper in this book is made from low- or no-chlorine pulp and is acid free, in conformance with international standards for paper permanency.

Lecture Notes in Artificial Intelligence (LNAI)

Lecture Notes in Computer Science